Lecture Notes in Mathematics

Edited by A. Dold and B. Eckmann
Series: Mathematisches Institut der Universität Bonn
Adviser: F. Hirzebruch

716

M. Scheunert

The Theory of Lie Superalgebras

An Introduction

Springer-Verlag
Berlin Heidelberg New York 1979

Author

Manfred Scheunert
Department of Physics
University of Wuppertal
D-5600 Wuppertal 1

AMS Subject Classifications (1970): 17 E05

ISBN 3-540-09256-0 Springer-Verlag Berlin Heidelberg New York
ISBN 0-387-09256-0 Springer-Verlag New York Heidelberg Berlin

Library of Congress Cataloging in Publication Data
Scheunert, M 1939- The theory of Lie superalgebras. (Lecture notes in mathematics; 716)
Includes bibliographical references and index. 1. Lie algebras. I. Title. II. Series: Lecture notes in mathematics (Berlin) ; 716.
QA3.L28 no. 716 [QA252.3] 510'.8s [512'.55] 79-15333

Printing and binding: Beltz Offsetdruck, Hemsbach/Bergstr.
2141/3140-543210

To Irene

PREFACE

The theory of Lie superalgebras (or, as they are also called, Z_2-graded Lie algebras) has undergone a remarkable evolution during the last few years. At present the most important result in the theory seems to be the classification by V.G. Kac of the finite-dimensional simple Lie superalgebras over an algebraically closed field of characteristic zero. Our main objective is to give a self-contained and detailed presentation of this classification. Thus we shall not presuppose any knowledge of the theory of Lie superalgebras, however, we assume that the reader is familiar with the standard theory of Lie algebras.

The present article has been written during the author's visit to the Dublin Institute for Advanced Studies, a stay which has been made possible through a grant by the Deutsche Forschungsgemeinschaft. The kind hospitality at DIAS as well as the support by the DFG are gratefully acknowledged. Above all, thanks are due to V. Rittenberg; without his permanent interest and encouragement this work would have hardly been written.

Dublin
April, 1978

Manfred Scheunert

TABLE OF CONTENTS

INTRODUCTION

During the last few years the theory of Lie superalgebras has seen a remarkable evolution, both in mathematics and physics. The reader who is interested in the historical background is referred to the review by Corwin, Ne'eman and Sternberg [1] which presents the subject as it was known in 1974. As a recent survey of the physical applications we mention the article by Fayet and Ferrara [2]. Both of these works contain an extensive bibliography. The most comprehensive description of the mathematical theory of Lie superalgebras is due to Kac [3] (a sketch of this article has been given in [4]).

The present work, too, is concerned with the mathematical side of the subject. Our main intent is to give a self-contained and detailed presentation of the classification of all finite-dimensional simple Lie superalgebras over an algebraically closed field of characteristic zero, a classification which has been obtained by Kac [3]. The difficulty lies in the fact that the Killing form of a simple Lie superalgebra may be equal to zero. Thus the techniques which are commonly used in the classification of semi-simple Lie algebras are not applicable here. But even if one is willing to assume in addition that the Killing form is non-degenerate one still has to cope with the problem that normally this form induces a non-definite bilinear form on the real vector space spanned by the roots. An investigation along these lines has been carried out by Kaplansky [5] (see also [3]).

Having in mind the classification of all simple Lie superalgebras we have to look for different techniques and only use the Killing form where it is already known to be non-degenerate (or else to exploit the very fact that the Killing form is equal to zero).

Let us describe the approach which will be chosen in the present work. To do so we have to be a little more explicit. A Lie superalgebra L is a Z_2 - graded algebra $L = L_{\bar{0}} \oplus L_{\bar{1}}$; $Z_2 = \{\bar{0},\bar{1}\}$, whose defining commutator identities involve signatures depending on the degrees of the elements. In particular, $L_{\bar{0}}$ is a Lie algebra and $L_{\bar{1}}$ is an $L_{\bar{0}}$ - module.

According to Kac the classification of simple Lie superalgebras is divided into two main parts. In the first part we give the classification of the so-called classical simple Lie superalgebras. A simple Lie superalgebra $L = L_{\bar{0}} \oplus L_{\bar{1}}$ is called classical if the representation of $L_{\bar{0}}$ in $L_{\bar{1}}$ is completely reducible. Remarkably enough it is exactly this class of Lie superalgebras to which the author and his co-workers were led from the physical side [6-8]. We have shown [6] that a simple Lie superalgebra $L = L_{\bar{0}} \oplus L_{\bar{1}}$ is classical if and only if the Lie algebra $L_{\bar{0}}$ is reductive. Now the reductive Lie algebras are those which are commonly used to describe internal symmetries of elementary particles. Hence the classical simple Lie superalgebras seemed to be a reasonable family of algebras to be classified.

In the present work we shall deal with the classical simple Lie superalgebras by means of the techniques which have been developed in [7,8] These have the advantage of starting directly from the Jacobi identity. In particular, they make quite evident "why" the two main double sequences of classical simple Lie superalgebras (the special linear and the orthosymplectic Lie superalgebras) exist. Regrettably, this method is not powerful enough to enable the classification of all classical simple Lie superalgebras, however, the remaining cases can be settled with a rather small amount of technicalities.

In the second part we obtain the classification of the non-classical simple Lie superalgebras. Here the filtrations and Z-gradations of Lie superalgebras play an important role. As an intermediate step we have to classify a certain family of transitive Z-graded Lie superalgebras which arise from filtrations of simple Lie superalgebras. The reasoning in this part is essentially due to Kac [3]. However, we have added several details and at some places simplified his arguments. In particular, we have avoided to use his theory of contragredient Lie superalgebras. Certainly, this theory is interesting in itself, but in our opinion it is not necessary (perhaps not even advantageous) to apply it in the proof of the classification theorem.

Our proof of the classification theorem appears to be somewhat incoherent in that we have to discuss various special cases. The reader might suspect that this is due to the fact that we have combined two different approaches to the subject. However, a short look at Kac's proof shows that this is not the case. Thus it is still worth-while to seek for some organizing principle which, finally, might allow of a uniform proof of the theorem.

Let us now give a brief account of the contents of this work. Chapter 0 is preparatory; we introduce our main conventions in §1 and make some general remarks on graded algebraic structures in §2.

Chapter I is formal in character. We give the basic definitions in §1, discuss the enveloping algebra of a Lie superalgebra in §2, and describe the usual elementary constructions with representations of Lie superalgebras in §3. Not all the material which is covered in §3 is really necessary for the rest of this work, however, it might be useful to have these constructions collected at some place. In §4 we introduce the concepts of induced and produced representations of Lie superalgebras and generalize the Guillemin, Sternberg theorem. This theorem will be important for the discussion of the non-classical simple Lie superalgebras.

Chapter II is devoted to the discussion of the simple Lie superalgebras and to the proof of the classification theorem. §1 is introductory; it provides some information on Z - gradations and filtrations of Lie superalgebras. In §2 we derive a few elementary properties of simple Lie superalgebras and prove the characterization of the classical simple Lie superalgebras which has been mentioned above. §3 contains several results on Lie superalgebras whose Killing form is non-degenerate. The next two paragraphs are devoted to the classical simple Lie superalgebras; they are described in §4 and classified in §5. The latter paragraph also contains some partial results on Z - graded Lie superalgebras. In §6 we give a detailed discussion of the so-called Cartan Lie superalgebras. (Let us remark that Kac has used the language of differential forms for describing these algebras.) §7 is devoted to the proof of one further partial result on Z - graded Lie superalgebras. In §8, at last, we are ready to classify the Z - graded Lie superalgebras which arise from certain filtrations of simple Lie superalgebras. The classification of the simple Lie superalgebras themselves is an easy consequence of the earlier results of this chapter.

In chapter III we give (without proofs) a survey of various further developments. §1 contains some results on superderivations of Clifford algebras and of Lie superalgebras, in §2 we make a few remarks on nilpotent, solvable and semi-simple Lie superalgebras, finally, in §3 we comment on the finite-dimensional representations of simple Lie superalgebras.

In general we shall presuppose a working knowledge of the theory of Lie algebras. A standard reference is Bourbaki's treatise [9 - 12]; our notational conventions as well as some special results are included in the appendix.

In the present work we shall not comment on the theory of supermanifolds or of Lie supergroups. The reader who is interested in these topics is referred to the literature [13 - 17].

Chapter 0 Preparatory remarks

§ 1 Conventions

1) In the present work we are dealing exclusively with vector spaces and algebras over a (commutative) field K of *characteristic zero*. All additional assumptions on the vector spaces and algebras (for example : finite-dimensional) or on the field K (for example : algebraically closed) will be mentioned explicitly; sometimes this will be done once and for all at the beginning of a paragraph or section.

2) An algebra over K is by definition a vector space (over K) equipped with some distributive (i.e. bilinear) multiplication. All additional assumptions on this multiplication (for example : associativity) will be mentioned explicitly.

3) The associative algebras appearing in this work will always contain a unit element.

A homomorphism of an associative algebra A into an associative algebra B is always assumed to map the unit element of A onto the unit element of B.

Every module M over an associative algebra A is assumed to be unitary (i.e. the multiplication by the unit element of A is the identity mapping of M).

4) Our notational conventions on Lie algebras are collected in the appendix.

§ 2 Some general remarks on graded algebraic structures

In this paragraph we recall the definitions of certain graded algebraic structures. For a more detailed discussion of such objects we refer the reader to the literature (for example, see [18 , 19]).

Let Γ be one of the rings Z (ring of integers) or $Z_2 = Z/2Z$ (ring of integers modulo 2). We shall only consider gradations with values in Γ. The two elements of Z_2 will be denoted by $\bar{0}$ (residue class of even integers) and $\bar{1}$ (residue class of odd integers). If $a \in Z$ then the integer $(-1)^a$ depends only on the residue class modulo 2 of a. Hence for all $\alpha \in Z_2$ the integer $(-1)^\alpha$ is well-defined.

1) Let V be a vector space over the field K. A Γ-*gradation of the vector space* V is a family $(V_\gamma)_{\gamma\in\Gamma}$ of subspaces of V such that

$$V = \bigoplus_{\gamma\in\Gamma} V_\gamma \quad . \tag{2.1}$$

The *vector space* V is said to be Γ-*graded* if it is equipped with a Γ-gradation.

An element of V is called *homogeneous* of degree γ, $\gamma \in \Gamma$, if it is an element of V_γ. In the case $\Gamma = Z_2$ the elements of $V_{\bar{0}}$ (resp. $V_{\bar{1}}$) are also called *even* (resp. *odd*).

Every element $y \in V$ has a unique decomposition of the form

$$y = \sum_{\gamma\in\Gamma} y_\gamma \quad ; \quad y_\gamma \in V_\gamma \ , \quad \gamma \in \Gamma \tag{2.2}$$

(where, of course, only finitely many y_γ are different from zero). The element y_γ is called the *homogeneous component* of y of degree γ.

A subspace U of V is called Γ-graded (or simply graded) if it contains the homogeneous components of all of its elements, i.e. if

$$U = \bigoplus_{\gamma\in\Gamma} (U \cap V_\gamma) \quad . \tag{2.3}$$

On any Z-graded vector space $V = \bigoplus_{j\in Z} V_j$ there exists a natural Z_2-gradation which is said to be induced by the Z - gradation and which is defined by

$$V_{\bar{0}} = \bigoplus_{j\in Z} V_{2j} \quad ; \quad V_{\bar{1}} = \bigoplus_{j\in Z} V_{2j+1} \quad . \tag{2.4}$$

2) Let

$$W = \bigoplus_{\gamma\in\Gamma} W_{\gamma} \tag{2.5}$$

be a second Γ - graded vector space. A linear mapping

$$g : V \longrightarrow W \tag{2.6}$$

is said to be *homogeneous* of degree γ , $\gamma \in \Gamma$, if

$$g(V_{\alpha}) \subset W_{\alpha+\gamma} \quad \text{for all } \alpha \in \Gamma \quad . \tag{2.7}$$

The mapping g is called a homomorphism of the Γ - graded vector space V into the Γ - graded vector space W if g is homogeneous of degree 0 . It is now evident how we define an isomorphism or an automorphism of Γ - graded vector spaces.

3) Let U and U' be two Γ - graded vector spaces over K . Then the tensor product $U \otimes U'$ has a natural Γ - gradation which is defined by

$$(U \otimes U')_{\gamma} = \bigoplus_{\alpha+\beta=\gamma} (U_{\alpha} \otimes U'_{\beta}) \quad , \quad \gamma \in \Gamma \ . \tag{2.8}$$

Suppose that we are given in addition two Γ - graded vector spaces V and V' . Let $g : U \longrightarrow V$ and $g' : U' \longrightarrow V'$ be two linear mappings which are homogeneous of degrees γ and γ', respectively. We define a linear mapping

$$g \mathbin{\bar{\otimes}} g' : U \otimes U' \longrightarrow V \otimes V' \tag{2.9,a}$$

by the requirement that

$$(g \mathbin{\bar{\otimes}} g')(x \otimes x') = (-1)^{\gamma'\xi} g(x) \otimes g'(x') \tag{2.9,b}$$

$$\text{for all } x \in U_{\xi} , \ x' \in U' \ ; \ \xi \in \Gamma \ .$$

Evidently, $g \mathbin{\bar{\otimes}} g'$ is a homogeneous linear mapping of degree $\gamma + \gamma'$.

If we are given two more Γ-graded vector spaces W and W' as well as two linear mappings $h : V \longrightarrow W$ and $h' : V' \longrightarrow W'$ which are homogeneous of degrees δ and δ', respectively, then we have

$$(h \bar{\otimes} h') \circ (g \bar{\otimes} g') = (-1)^{\delta'\gamma} (h \circ g) \bar{\otimes} (h' \circ g') . \qquad (2.10)$$

Remark 1)

Let $\langle \, , \, \rangle$ denote the "super-commutator" as defined in chapter I, §1, example 2). Suppose that U = V = W and U' = V' = W'; if $\langle h , g \rangle = 0$ and $\langle h', g' \rangle = 0$ it follows that $\langle h \bar{\otimes} h', g \bar{\otimes} g' \rangle = 0$, too.

It is obvious how to generalize all these results to tensor products of finitely many Γ-graded vector spaces.

4) Let A be an algebra over K. The *algebra* A is said to be Γ-*graded* if the underlying vector space of A is Γ-graded,

$$A = \bigoplus_{\gamma \in \Gamma} A_\gamma , \qquad (2.11)$$

and if, furthermore,

$$A_\alpha A_\beta \subset A_{\alpha+\beta} \quad \text{for all } \alpha, \beta \in \Gamma . \qquad (2.12)$$

Evidently, A_0 is a subalgebra of A. If A has a unit element then this element lies in A_0.

A homomorphism of Γ-graded algebras is by definition a homomorphism of the underlying algebras as well as of the underlying Γ-graded vector spaces; in particular, a homomorphism is homogeneous of degree 0. Similar remarks apply for isomorphisms and automorphisms.

A graded subalgebra (resp. ideal) of a Γ-graded algebra A is a subalgebra (resp. ideal) of the algebra A which is, in addition, a graded subspace of the Γ-graded vector space A. The quotient algebra of a Γ-graded algebra modulo a (two-sided) graded ideal is again a Γ-graded algebra.

If A and B are two Γ-graded algebras the direct (i.e. the cartesian) product $A \times B$ is an algebra which becomes a Γ-graded algebra by means of the definition

$$(A \times B)_\gamma = A_\gamma \times B_\gamma \quad \text{for all } \gamma \in \Gamma \ . \tag{2.13}$$

5) Let A be a Z - graded algebra and let A' be the Z - graded algebra whose underlying algebra is equal to that of A but whose Z - gradation is given by

$$A'_j = A_{-j} \quad \text{for all } j \in Z \ . \tag{2.14}$$

Then A' is called the Z - graded algebra obtained from A by inversion of the Z - gradation. Note that according to our definitions the Z - graded algebras A and A' are not necessarily isomorphic.

6) Let A and B be two Γ - graded associative algebras (recall that according to our conventions this implies that A and B have a unit element). On the Γ - graded vector space $A \otimes B$ (see 3)) we define a multiplication by the requirement that

$$(a \otimes b)(a' \otimes b') = (-1)^{\beta\alpha'}(a a') \otimes (b b') \tag{2.15}$$

$$\text{for all } a \in A \, , \ b \in B_\beta \, , \ a' \in A_{\alpha'} \, , \ b' \in B \ ; \ \beta \, , \alpha' \in \Gamma \ .$$

It is easy to check that with this multiplication $A \otimes B$ is a Γ - graded associative algebra [19]. This algebra will be called the *graded tensor product* of the Γ - graded algebras A and B and will be denoted by $A \mathbin{\bar{\otimes}} B$ (in order to avoid a confusion with the more usual definition).

The algebras $A \mathbin{\bar{\otimes}} B$ and $B \mathbin{\bar{\otimes}} A$ are canonically isomorphic. In fact, it is easy to see that there exists a unique linear mapping

$$s : A \mathbin{\bar{\otimes}} B \longrightarrow B \mathbin{\bar{\otimes}} A \tag{2.16,a}$$

such that

$$s(a \otimes b) = (-1)^{\alpha\beta}\, b \otimes a \tag{2.16,b}$$

$$\text{for all } a \in A_\alpha \, , \ b \in B_\beta \ ; \ \alpha \, , \beta \in \Gamma$$

and that this mapping is an isomorphism of Γ - graded algebras.

The definition of the graded tensor product of associative Γ - graded algebras is easily generalized to the case of more than two factors [19];

this construction is still "associative" in the usual sense.

7) Let A be a Γ - graded associative algebra and let V be a left A - module (recall that according to our conventions A has a unit element and that V is unitary). In particular, V is a vector space over K . The A - *module* V is said to be Γ - *graded* if the underlying vector space of V is Γ - graded and if, furthermore,

$$A_\alpha V_\beta \subset V_{\alpha+\beta} \quad \text{for all } \alpha , \beta \in \Gamma \quad . \tag{2.17}$$

Right Γ - graded A - modules are defined similarly.

A homomorphism of Γ - graded A - modules is by definition a homomorphism of the underlying A - modules as well as of the Γ - graded vector spaces (i.e. it is A - linear and homogeneous of degree 0). Similar remarks apply for isomorphisms and automorphisms.

8) Let A and B be two Γ - graded associative algebras and let V (resp. W) be a Γ - graded left A - module (resp. B - module). Then $V \otimes W$ is a Γ - graded vector space and $A \mathbin{\bar{\otimes}} B$ is a Γ - graded associative algebra (see 3) and 6)). It is easy to see that there exists a unique structure of a left Γ - graded $A \mathbin{\bar{\otimes}} B$ - module on $V \otimes W$ such that

$$(a \otimes b)(x \otimes y) = (-1)^{\beta\xi} (a x) \otimes (b y) \tag{2.18}$$

for all $a \in A$, $b \in B_\beta$, $x \in V_\xi$, $y \in W$; $\beta , \xi \in \Gamma$.

9) Finally, we shall introduce some notions for the case where an algebra is equipped with both a Z_2 - gradation and a Z - gradation. It is convenient to define [3] :

Definition 1

A Z_2 - graded algebra is called a *superalgebra*.

Definition 2

A *superalgebra* S is said to be Z - *graded* if we are given a family $(S_j)_{j\in Z}$ of Z_2 - graded subspaces of S such that

$$S = \bigoplus_{j \in Z} S_j \qquad (2.19,a)$$

$$S_i S_j \subset S_{i+j} \quad \text{for all } i, j \in Z \ . \qquad (2.19,b)$$

The Z - gradation $(S_j)_{j \in Z}$ is said to be *consistent* with the Z_2 - gradation of S if

$$S_{\bar{0}} = \bigoplus_{j \in Z} S_{2j} \quad ; \quad S_{\bar{1}} = \bigoplus_{j \in Z} S_{2j+1} \ . \qquad (2.20)$$

According to this definition a Z - graded superalgebra is just a Δ - graded algebra, Δ being the additive group $Z \times Z_2$. Furthermore, a superalgebra with a consistent Z - gradation (or, as we shall say, a consistently Z - graded superalgebra) is nothing but a Z - graded algebra which is equipped in addition with the Z_2 - gradation induced by its Z - gradation.

CHAPTER I FORMAL CONSTRUCTIONS

§ 1 DEFINITION AND ELEMENTARY PROPERTIES OF LIE SUPERALGEBRAS

Recall (chapter 0, definition 1) that a superalgebra is by definition nothing else but a Z_2 - graded algebra.

Definition 1

Let $L = L_{\bar{0}} \oplus L_{\bar{1}}$ be a superalgebra whose multiplication is denoted by a pointed bracket $\langle \, , \, \rangle$. This implies in particular that

$$\langle L_\alpha , L_\beta \rangle \subset L_{\alpha+\beta} \quad \text{for all } \alpha , \beta \in Z_2 \; . \tag{1.1}$$

We call L a *Lie superalgebra* if the multiplication satisfies the following identities :

$$\langle A , B \rangle = -(-1)^{\alpha\beta} \langle B , A \rangle \tag{1.2}$$

(graded skew-symmetry)

$$(-1)^{\gamma\alpha} \langle A , \langle B , C \rangle\rangle + (-1)^{\alpha\beta} \langle B , \langle C , A \rangle\rangle + (-1)^{\beta\gamma} \langle C , \langle A , B \rangle\rangle = 0$$

(graded Jacobi identity) (1.3)

for all $A \in L_\alpha$, $B \in L_\beta$, $C \in L_\gamma$; $\alpha , \beta , \gamma \in Z_2$.

Remark 1)

Lie superalgebras are frequently called Z_2 - graded Lie algebras in spite of the fact that in general they are *not* Lie algebras.

Let L be a Lie superalgebra. All the following statements are obvious.

a) The subalgebra $L_{\bar{0}}$ of L is a Lie algebra.

b) If the graded vector space L is equipped with the "inverted multiplication" $(A , B) \longrightarrow \langle B , A \rangle$ we obtain again a Lie superalgebra.

c) The definitions of a graded subalgebra, a graded ideal, a graded quotient algebra of L are standard and need not be repeated here (see chapter 0, §2, 4)). Let us note, however, that a left (or right) graded ideal of L is automatically a two-sided graded ideal (this is no longer true if we drop the requirement that the ideal should be graded).

Example 1)

The so-called commutator algebra $\langle L , L \rangle$ of L is a graded ideal of L [20]

d) The direct product of two Lie superalgebras is a Lie superalgebra.

e) Let E be a field containing K. The algebra $E \underset{K}{\otimes} L$ obtained from L by extension of the base field from K to E has a natural Z_2 - gradation which is defined by

$$(E \underset{K}{\otimes} L)_{\alpha} = E \underset{K}{\otimes} L_{\alpha} \quad \text{for all } \alpha \in Z_2 \quad . \tag{1.4}$$

It is obvious that $E \underset{K}{\otimes} L$ is a Lie superalgebra over E.

Finally, let us recall that *homomorphisms (isomorphisms, automorphisms) of Lie superalgebras are always assumed to be consistent with the* Z_2 *- gradations*, i.e. they are homogeneous linear mappings of degree zero.

Examples

In chapter II we shall describe a large multitude of various Lie superalgebras. Hence we shall restrict our attention to some examples which are useful for the development of the general theory.

2) Let $S = S_{\bar{0}} \oplus S_{\bar{1}}$ be an associative superalgebra. On the Z_2 - graded vector space S we define a new multiplication $\langle \ , \ \rangle$ by

$$\langle A , B \rangle = A B - (-1)^{\alpha\beta} B A \tag{1.5}$$

$$\text{for all } A \in S_{\alpha} , \ B \in S_{\beta} ; \ \alpha , \beta \in Z_2 \quad .$$

The algebra which emerges will be denoted by $\tilde{S}$; it is easily seen to be a Lie superalgebra and is said to be associated with the associative superalgebra S. This example makes evident in which sense the multiplication in a Lie superalgebra behaves "partly as a commutator and partly as an anticommutator".

3) An important special case of example 2) is the following :

Let $V = V_{\bar{0}} \oplus V_{\bar{1}}$ be a Z_2 - graded vector space. The algebra Hom(V) (consisting of the K - linear mappings of V into itself) becomes an associative superalgebra if one defines the Z_2 - gradation by

$$\mathrm{Hom}(V)_\alpha = \{ A \in \mathrm{Hom}(V) \mid A(V_\beta) \subset V_{\alpha+\beta} , \beta \in Z_2 \} \quad (1.6)$$

for all $\alpha \in Z_2$. (Hence $\mathrm{Hom}(V)_\alpha$ consists of the linear mappings of V into itself which are homogeneous of degree α.)

The Lie superalgebra associated with Hom(V) will be denoted by pl(V) and will be called the *general linear Lie superalgebra of* V ; it plays the same role in the theory of Lie superalgebras as the Lie algebra gl(V) does in the theory of (ordinary) Lie algebras.

In particular, we define :

Definition 2

A *graded representation* of a Lie superalgebra L in a Z_2 - graded vector space V is an (even) homomorphism of L into pl(V) .

4) Suppose that the vector space V of example 3) is a superalgebra $T = T_{\bar{0}} \oplus T_{\bar{1}}$. Let $\mathcal{D}_\alpha(T)$, $\alpha \in Z_2$, be the subspace of all $A \in pl_\alpha(T)$ such that

$$A(xy) = (Ax)y + (-1)^{\alpha\xi} x(Ay) \quad (1.7)$$

$$\text{for all } x \in T_\xi , \; y \in T ; \; \xi \in Z_2 .$$

Hence if $\alpha = \bar{0}$ (resp. $\alpha = \bar{1}$) the subspace $\mathcal{D}_\alpha(T)$ consists of the even derivations (resp. of the odd antiderivations) of the superalgebra T. We define

$$\mathcal{D}(T) = \mathcal{D}_{\bar{0}}(T) \oplus \mathcal{D}_{\bar{1}}(T) \; ; \quad (1.8)$$

then it is easy to see that $\mathcal{D}(T)$ is a graded subalgebra of pl(T) . The elements of $\mathcal{D}(T)$ are called *superderivations* of T , hence $\mathcal{D}(T)$ is called the *Lie superalgebra of superderivations of* T .

We shall now assume that T is associative. Let $\tilde{T}$ be the Lie superalgebra associated with T (see example 2)). For every $a \in T$ we define a linear map $\hat{a}$ of T into itself by

$$\hat{a}(x) = \langle a, x \rangle \quad \text{for all } x \in T \quad . \tag{1.9}$$

It is easy to see that $\hat{a} \in \mathfrak{D}(T)$ and that the mapping

$$\tilde{T} \longrightarrow \mathfrak{D}(T) \quad , \quad a \longrightarrow \hat{a} \tag{1.10}$$

is a homomorphism of Lie superalgebras. The superderivations of the form $\hat{a}$, $a \in T$, are called *inner*.

Later on we shall discuss the algebra $\mathfrak{D}(T)$ for the special cases in which T is an exterior algebra (see chapter II, §6), a Clifford algebra (see chapter III, §1, n°1) or a Lie superalgebra (see chapter III, §1, n°2).

Our examples lead to the following interpretation of the graded Jacobi identity (1.3). If $A \in L$ we define the linear mapping

$$\operatorname{ad} A : L \longrightarrow L \tag{1.11,a}$$

by

$$(\operatorname{ad} A)(B) = \langle A, B \rangle \quad \text{for all } B \in L \quad . \tag{1.11,b}$$

Taking the graded skew-symmetry (1.2) for granted the graded Jacobi identity is equivalent to each of the following requirements :

1) ad is a homomorphism of the superalgebra L into the Lie superalgebra pl(L) . This homomorphism is called the *adjoint representation* of L .

2) ad A is a superderivation of L , for all $A \in L$.

Combining these two results we see that ad is a homomorphism of the Lie superalgebra L into the Lie superalgebra $\mathfrak{D}(L)$. The superderivations of L which are of the form ad A , $A \in L$, are called *inner*.

We have already mentioned above that $L_{\bar{0}}$ is a Lie algebra. Hence the restriction of ad to $L_{\bar{0}}$ is a representation of $L_{\bar{0}}$ in the vector space L . Since the subspaces $L_{\bar{0}}$ and $L_{\bar{1}}$ of L are invariant under this representation we can define :

Definition 3

The adjoint representation ad of the Lie superalgebra L induces a representation of the Lie algebra $L_{\bar{0}}$ in the odd subspace $L_{\bar{1}}$. This representation is called the *adjoint representation of* $L_{\bar{0}}$ *in* $L_{\bar{1}}$ and is denoted by ad'.

We now are ready to give a new description of Lie superalgebras, as follows.

Let L be a Lie superalgebra. For convenience we shall write for the moment $\tilde{Q}$ instead of ad'Q , $Q \in L_{\bar{0}}$, hence

$$\tilde{Q}(U) = \langle Q , U \rangle = -\langle U , Q \rangle \quad \text{for all } Q \in L_{\bar{0}},\ U \in L_{\bar{1}} . \tag{1.12}$$

The Lie superalgebra L is uniquely fixed if we are given the Lie algebra $L_{\bar{0}}$, the representation $Q \longrightarrow \tilde{Q}$ of $L_{\bar{0}}$ in $L_{\bar{1}}$ and the symmetric bilinear product mapping

$$P : L_{\bar{1}} \times L_{\bar{1}} \longrightarrow L_{\bar{0}} . \tag{1.13}$$

The graded Jacobi identity implies that P is $L_{\bar{0}}$-invariant, i.e. that

$$\langle Q , P(U,V) \rangle = P(\tilde{Q}(U),V) + P(U,\tilde{Q}(V)) \tag{1.14}$$
$$\text{for all } Q \in L_{\bar{0}} ;\ U , V \in L_{\bar{1}} ,$$

and that, furthermore,

$$P(U,V)^{\sim}(W) + P(V,W)^{\sim}(U) + P(W,U)^{\sim}(V) = 0 \tag{1.15}$$
$$\text{for all } U , V , W \in L_{\bar{1}} .$$

Conversely, let $L_{\bar{0}}$ be a Lie algebra and let $Q \longrightarrow \tilde{Q}$ be a representation of $L_{\bar{0}}$ in some vector space $L_{\bar{1}}$. Suppose we are given a symmetric bilinear mapping P of $L_{\bar{1}} \times L_{\bar{1}}$ into $L_{\bar{0}}$. We define a multiplication $\langle\ ,\ \rangle$ on the Z_2-graded vector space $L_{\bar{0}} \oplus L_{\bar{1}}$ by

$$\langle Q , R \rangle = [Q , R] \qquad \text{if } Q , R \in L_{\bar{0}} \tag{1.16,a}$$
$$\langle Q , U \rangle = -\langle U , Q \rangle = \tilde{Q}(U) \quad \text{if } Q \in L_{\bar{0}},\ U \in L_{\bar{1}} \tag{1.16,b}$$
$$\langle U , V \rangle = P(U,V) \qquad \text{if } U , V \in L_{\bar{1}} . \tag{1.16,c}$$

With this multiplication $L_{\bar{0}} \oplus L_{\bar{1}}$ is a Lie superalgebra if and only if P is $L_{\bar{0}}$ - invariant and if (1.15) is fulfilled.

In a sense, therefore, a Lie superalgebra $L = L_{\bar{0}} \oplus L_{\bar{1}}$ is some "superstructure" to be built over the Lie algebra $L_{\bar{0}}$. This point of view is quite useful for both the general theory and the explicit construction of Lie superalgebras.

Example 5)

Choose three 2 - dimensional vector spaces V_i , $i = 1,2,3$. Let ψ_i be a non-degenerate skew-symmetric bilinear form on V_i (any two such forms are proportional). It is well-known that $sp(\psi_i) = sl(V_i)$. We define

$$L_{\bar{0}} = sp(\psi_1) \times sp(\psi_2) \times sp(\psi_3) \quad , \quad L_{\bar{1}} = V_1 \otimes V_2 \otimes V_3 \quad . \tag{1.17}$$

As a representation of $L_{\bar{0}}$ in $L_{\bar{1}}$ we choose the tensor product of the natural representations of $sp(\psi_i)$ in V_i .

Up to a factor there exists just one $sp(\psi_i)$ - invariant bilinear mapping

$$P_i : V_i \times V_i \longrightarrow sp(\psi_i) \quad ; \tag{1.18,a}$$

we choose the normalization such that

$$P_i(x_i,y_i)\, z_i = \psi_i(y_i,z_i)\, x_i - \psi_i(z_i,x_i)\, y_i \tag{1.18,b}$$
$$\text{for all } x_i , y_i , z_i \in V_i \quad .$$

Then the most general ansatz for a $L_{\bar{0}}$ - invariant bilinear mapping

$$P : L_{\bar{1}} \times L_{\bar{1}} \longrightarrow L_{\bar{0}} \tag{1.19,a}$$

is given by

$$\begin{aligned} P(\, x_1 \otimes x_2 \otimes x_3 , y_1 \otimes y_2 \otimes y_3 \,) &= \sigma_1\, \psi_2(x_2,y_2)\, \psi_3(x_3,y_3)\, P_1(x_1,y_1) \\ &+ \sigma_2\, \psi_1(x_1,y_1)\, \psi_3(x_3,y_3)\, P_2(x_2,y_2) \\ &+ \sigma_3\, \psi_1(x_1,y_1)\, \psi_2(x_2,y_2)\, P_3(x_3,y_3) \end{aligned} \tag{1.19,b}$$
$$\text{for all } x_i , y_i \in V_i \; ; \; i = 1,2,3 \quad .$$

Here $\sigma_1, \sigma_2, \sigma_3$ are some arbitrary elements of K. The mapping P is symmetric and it is easy to check that the identity (1.15) is fulfilled if and only if

$$\sigma_1 + \sigma_2 + \sigma_3 = 0 \quad . \tag{1.20}$$

If this condition is satisfied, then $L = L_{\bar{0}} \oplus L_{\bar{1}}$ is a Lie superalgebra which will be denoted by $\Gamma(\sigma_1,\sigma_2,\sigma_3)$. The following remark shows that this notation is adequate. Suppose we are given, for $i = 1,2,3$, a 2-dimensional vector space V_i', a non-degenerate skew-symmetric bilinear form ψ_i' on V_i', and a constant $\sigma_i' \in K$. We assume that $\sigma_1' + \sigma_2' + \sigma_3' = 0$ and apply the above construction to obtain a Lie superalgebra L'. Then the Lie superalgebras L and L' are isomorphic if and only if there exists a permutation π of $\{1,2,3\}$ and a non-zero element $\tau \in K$ such that

$$\sigma_i' = \tau \cdot \sigma_{\pi i} \quad \text{for } i = 1,2,3 \quad . \tag{1.21}$$

§ 2 The enveloping algebra of a Lie superalgebra

In this paragraph we shall introduce the enveloping algebra of a Lie superalgebra and describe some of its properties. Our discussion will be completely analogous to the corresponding one for Lie algebras [9, 10]. Hence we may skip all the proofs (apart from the proof of the Poincaré, Birkhoff, Witt theorem these proofs are obvious anyway). As is to be expected from the Lie algebra case the enveloping algebra turns out to be a very useful tool for the theory of Lie superalgebras and of their representations.

1. Definition and some basic properties of the enveloping algebra

Let $L = L_{\bar{0}} \oplus L_{\bar{1}}$ be a Lie superalgebra and let $T(L)$ be the tensor algebra of the vector space L. The Z_2 - gradation of L induces a Z_2 - gradation of $T(L)$ such that the canonical injection $L \longrightarrow T(L)$ is an even linear mapping and that $T(L)$ is a superalgebra (see chapter 0, §2, 3)). We consider the two-sided ideal J of $T(L)$ which is generated by the elements of the form

$$A \otimes B - (-1)^{\alpha\beta} B \otimes A - \langle A , B \rangle \tag{2.1}$$

$$\text{with } A \in L_\alpha , \; B \in L_\beta ; \; \alpha , \beta \in Z_2 .$$

Evidently these elements are homogeneous (of degree $\alpha + \beta$), hence J is a graded ideal. Therefore, if we define

$$U(L) = T(L)/J , \tag{2.2}$$

it follows that $U(L)$ is an associative superalgebra; this algebra is called the *(universal) enveloping algebra of* L. By composing the canonical injection $L \longrightarrow T(L)$ with the canonical mapping $T(L) \longrightarrow T(L)/J = U(L)$ we obtain the canonical even linear mapping

$$\sigma : L \longrightarrow U(L) \tag{2.3}$$

which satisfies the following condition :

$$\sigma(\langle A, B\rangle) = \sigma(A)\,\sigma(B) - (-1)^{\alpha\beta}\,\sigma(B)\,\sigma(A) \tag{2.4}$$

$$\text{for all } A \in L_\alpha,\ B \in L_\beta\ ;\ \alpha, \beta \in Z_2\ .$$

Every element of U(L) is a linear combination of products of the form

$$\sigma(A_1) \ldots\ldots \sigma(A_r) \quad \text{with } A_i \in L_{\alpha_i},\ \alpha_i \in Z_2\ ;\ 1 \leq i \leq r \tag{2.5}$$

(for r = 0 we define this product to be equal to 1); the product in (2.5) is a homogeneous element of U(L) of degree $\alpha_1 + \ldots. + \alpha_r$.

The pair (U(L),σ) is characterized by (2.4) and by the following universal property :

Proposition 1

Let S be an associative algebra with unit element and let g be a linear mapping of L into S such that

$$g(\langle A, B\rangle) = g(A)\,g(B) - (-1)^{\alpha\beta}\,g(B)\,g(A) \tag{2.6}$$

$$\text{for all } A \in L_\alpha,\ B \in L_\beta\ ;\ \alpha, \beta \in Z_2\ .$$

Then there exists a unique homomorphism $\bar{g}$ of the algebra U(L) into the algebra S such that

$$g = \bar{g} \circ \sigma \quad , \quad \bar{g}(1) = 1\ . \tag{2.7}$$

Corollary 1

If S is a superalgebra and if g is homogeneous of degree zero then $\bar{g}$ is homogeneous of degree zero, too (i.e. $\bar{g}$ is a homomorphism of superalgebras).

Corollary 2

Let L and L' be two Lie superalgebras and let σ (resp. σ') be the canonical mapping of L into U(L) (resp. of L' into U(L')). If $h : L \longrightarrow L'$ is a homomorphism of Lie superalgebras there exists a unique homomorphism $\bar{h} : U(L) \longrightarrow U(L')$ of superalgebras such that

$$\sigma' \circ h = \bar{h} \circ \sigma \quad , \quad \bar{h}(1) = 1\ . \tag{2.8}$$

Let L" be a third Lie superalgebra and let σ" be the canonical mapping of L" into U(L"). If $h' : L' \longrightarrow L''$ is a homomorphism of Lie superalgebras we have

$$\overline{h' \circ h} = \overline{h'} \circ \overline{h} \quad . \tag{2.9}$$

Furthermore,

$$\overline{id_L} = id_{U(L)} \quad . \tag{2.10}$$

We proceed by describing some important properties of the pair (U(L),σ). First we remark that there exists a "graded counterpart" to the "principal anti-automorphism" of the enveloping algebra of a Lie algebra [9]. In fact, it is easy to construct an even linear mapping

$$\theta : U(L) \longrightarrow U(L) \tag{2.11,a}$$

such that

$$\theta(XY) = (-1)^{\xi\eta}\,\theta(Y)\,\theta(X) \tag{2.11,b}$$

for all $X \in U(L)_\xi$, $Y \in U(L)_\eta$; $\xi, \eta \in Z_2$

$$\theta(\sigma(A)) = -\sigma(A) \tag{2.11,c}$$

for all $A \in L$

$$\theta(1) = 1 \tag{2.11,d}$$

and from these properties we deduce that

$$\theta^2 = id \quad . \tag{2.11,e}$$

To give an example of how the proposition 1 and its corollaries may be applied we discuss in some detail the enveloping algebra of the direct product L × L' of two Lie superalgebras L and L'. The canonical mappings of L into U(L) and of L' into U(L') will be denoted by σ and σ', respectively.

Let S be an associative algebra (with unit element) and let $g : L \times L' \rightarrow S$ be a linear mapping such that the condition (2.6) is fulfilled. Let h (resp. h') be the restriction of g onto L (resp. onto L'). Then (2.6) is valid for h and h', too. Hence there exist two algebra - homomorphisms

$$\overline{h} : U(L) \longrightarrow S \quad , \quad \overline{h'} : U(L') \longrightarrow S \tag{2.12}$$

such that

$$h = \overline{h} \circ \sigma \quad , \quad \overline{h}(1) = 1 \tag{2.13}$$

$$h' = \overline{h'} \circ \sigma' \quad , \quad \overline{h'}(1) = 1 \ . \tag{2.13'}$$

It is easy to see that

$$\overline{h}(X)\, \overline{h'}(X') = (-1)^{\xi\xi'}\, \overline{h'}(X')\, \overline{h}(X) \tag{2.14}$$

$$\text{for all } X \in U(L)_{\xi} \, , \; X' \in U(L')_{\xi'} \; ; \; \xi , \xi' \in Z_2 \ .$$

Hence, if $U(L) \mathbin{\overline{\otimes}} U(L')$ is the graded tensor product of the superalgebras U(L) and U(L') (see chapter 0, §2, 6)), there exists an algebra-homomorphism

$$\hat{g} : U(L) \mathbin{\overline{\otimes}} U(L') \longrightarrow S \tag{2.15,a}$$

such that

$$\hat{g}(X \otimes X') = \overline{h}(X)\, \overline{h'}(X') \tag{2.15,b}$$

$$\text{for all } X \in U(L) \, , \; X' \in U(L') \ .$$

Now we define a linear mapping

$$\tau : L \times L' \longrightarrow U(L) \mathbin{\overline{\otimes}} U(L') \tag{2.16,a}$$

by

$$\tau(A, A') = \sigma(A) \otimes 1 + 1 \otimes \sigma'(A') \tag{2.16,b}$$

$$\text{for all } A \in L \, , \; A' \in L' \ .$$

Then τ satisfies the condition (2.4) and

$$g = \hat{g} \circ \tau \ . \tag{2.17}$$

We conclude :

<u>Proposition 2</u>

The enveloping algebra $U(L \times L')$ of $L \times L'$ is canonically isomorphic to the graded tensor product $U(L) \bar{\otimes} U(L')$ and the mapping τ (as defined in (2.16)) corresponds to the canonical mapping of $L \times L'$ into $U(L \times L')$.

As another result of this type we mention the following. Let L be a Lie superalgebra and let σ be the canonical mapping of L into the enveloping algebra $U(L)$. Furthermore, let E be a field containing K. If $E \underset{K}{\otimes} L$ (resp. $E \underset{K}{\otimes} U(L)$) is the algebra obtained from L (resp. $U(L)$) by extension of the base field from K to E then $U(E \underset{K}{\otimes} L)$ is canonically isomorphic to $E \underset{K}{\otimes} U(L)$ and the mapping $id_E \otimes \sigma$ corresponds to the canonical mapping of $E \underset{K}{\otimes} L$ into $U(E \underset{K}{\otimes} L)$.

2. The supersymmetric algebra of a graded vector space

Let $V = V_{\bar{0}} \oplus V_{\bar{1}}$ be a Z_2-graded vector space and let $T(V)$ be the tensor algebra of V. As is well-known $T(V)$ is a Z-graded algebra,

$$T(V) = \bigoplus_{n \in Z} T_n(V) \quad , \tag{2.18}$$

where $T_m(V) = \{0\}$ if $m \leqslant -1$ and where $T_n(V)$ is the vector space of tensors of order n if $n \geqslant 0$. If $T(V)$ is equipped with the Z_2-gradation inherited from V (see n°1) then all the $T_n(V)$ are Z_2-graded subspaces of $T(V)$; hence $T(V)$ is an associative Z-graded superalgebra (see chapter 0, §2, definition 2).

Now let $\tilde{J}$ be the two-sided ideal of $T(V)$ which is generated by the tensors of the form

$$A \otimes B - (-1)^{\alpha\beta} B \otimes A \tag{2.19}$$

$$\text{with } A \in L_\alpha \,,\; B \in L_\beta \,;\; \alpha, \beta \in Z_2 \;.$$

These tensors are homogeneous both with respect to the Z_2-gradation and with respect to the Z-gradation of $T(V)$. Hence

$$\tilde{U}(V) = T(V)/\tilde{J} \tag{2.20}$$

is an associative Z-graded superalgebra which is called the *supersym-*

metric algebra of the Z_2 - graded vector space V .

Evidently, the multiplication in $\tilde{U}(V)$ is supercommutative (in the obvious sense). If V is equipped with the trivial multiplication

$$\langle V , V \rangle = \{0\} \tag{2.21}$$

then V is an (abelian) Lie superalgebra and $\tilde{U}(V)$ is nothing else but the enveloping algebra of this Lie superalgebra.

The structure of $\tilde{U}(V)$ is easily found out. According to proposition 2 the superalgebra $\tilde{U}(V)$ is isomorphic to the tensor product

$$P = S(V_{\bar{0}}) \otimes \bigwedge V_{\bar{1}} \tag{2.22}$$

of the symmetric algebra of $V_{\bar{0}}$ with the exterior algebra of $V_{\bar{1}}$; the Z_2 - gradation of P is given by

$$P_{\bar{0}} = S(V_{\bar{0}}) \otimes \Big(\bigoplus_{j \geqslant 0} \bigwedge^{2j} V_{\bar{1}} \Big) \tag{2.23,a}$$

$$P_{\bar{1}} = S(V_{\bar{0}}) \otimes \Big(\bigoplus_{j \geqslant 0} \bigwedge^{2j+1} V_{\bar{1}} \Big) \quad . \tag{2.23,b}$$

On the other hand $S(V_{\bar{0}})$ and $\bigwedge V_{\bar{1}}$ are Z - graded algebras

$$S(V_{\bar{0}}) = \bigoplus_{n \in Z} S_n(V_{\bar{0}}) \tag{2.24}$$

$$\bigwedge V_{\bar{1}} = \bigoplus_{n \in Z} \bigwedge^{n} V_{\bar{1}} \quad , \tag{2.25}$$

the subspaces of degrees $n \leqslant -1$ being equal to $\{0\}$. It is obvious that the Z - gradation of $\tilde{U}(V)$ corresponds to the natural Z - gradation of the tensor product P : For $n \in Z$ we have

$$P_n = \{0\} \qquad \text{if } n \leqslant -1 \tag{2.26,a}$$

$$P_n = \bigoplus_{m=0}^{n} \Big(S_m(V_{\bar{0}}) \otimes \bigwedge^{n-m} V_{\bar{1}} \Big) \quad \text{if } n \geqslant 0 \quad . \tag{2.26,b}$$

3. Filtration of the enveloping algebra and the Poincaré, Birkhoff, Witt theorem

Let L be a Lie superalgebra and let T(L) be the tensor algebra of the vector space L . In the preceding section we have seen that T(L) has a natural structure of a Z - graded superalgebra

$$T(L) = \bigoplus_{n \in Z} T_n(L) \ . \tag{2.27}$$

We define for all $n \in Z$

$$T^n(L) = \bigoplus_{m \leq n} T_m(L) \ ; \tag{2.28}$$

the $T^n(L)$ are Z_2 - graded subspaces of T(L) .

Now let U(L) be the enveloping algebra of L and let $\sigma : L \longrightarrow U(L)$ be the canonical mapping. Recall that U(L) = T(L)/J where the Z_2 - graded ideal J of T(L) has been defined in n°1 of this paragraph. Let $U^n(L)$ be the image of $T^n(L)$ under the canonical mapping $T(L) \longrightarrow U(L)$. Then it is easy to verify the following statements :

a) If $n \geq 0$ the subspace $U^n(L)$ of U(L) is generated by the products of the form $\sigma(A_1) \ldots \sigma(A_m)$ with $0 \leq m \leq n$ and $A_1, \ldots, A_m \in L$.

b) The $U^n(L)$ are Z_2 - graded subspaces of U(L) .

c) $$U^n(L) \subset U^m(L) \quad \text{if } n \leq m \tag{2.29}$$

d) $$U^n(L) = \{0\} \quad \text{if } n \leq -1 \tag{2.30}$$

e) $$U^0(L) = K \cdot 1 \tag{2.31}$$

f) $$\bigcup_{n \geq 0} U^n(L) = U(L) \tag{2.32}$$

g) $$U^n(L)\, U^m(L) \subset U^{n+m}(L) \quad \text{for all } n, m \in Z \ . \tag{2.33}$$

The family $(U^n(L))_{n \in Z}$ is called the *canonical filtration* of the enveloping algebra U(L) .

Now let

$$G(L) \;=\; \bigoplus_{n\in Z} G_n(L) \qquad (2.34)$$

be the Z - graded algebra associated with the filtered algebra U(L) . Recall that

$$G_n(L) \;=\; U^n(L)\,/\,U^{n-1}(L) \qquad (2.35)$$

and that the multiplication in G(L) is obtained from that in U(L) by going to the quotients. Evidently, G(L) has a natural Z_2 - gradation, too; equipped with its two gradations G(L) is a Z - graded associative superalgebra.

Denote the composition of the canonical mappings $T_n(L) \longrightarrow U^n(L) \longrightarrow G_n(L)$ by φ_n and let $\varphi : T(L) \longrightarrow G(L)$ be the linear mapping defined by the family $(\varphi_n)_{n\in Z}$. Then φ is a surjective homomorphism of Z - graded superalgebras which vanishes on all tensors of the form

$$A \otimes B \;-\; (-1)^{\alpha\beta}\, B \otimes A \qquad (2.36)$$

$$\text{with } A \in L_\alpha\,,\; B \in L_\beta\,;\; \alpha\,,\beta \in Z_2 \;.$$

Consequently, φ defines a homomorphism $\tilde{\varphi}$ of the Z - graded superalgebra $\tilde{U}(L)$ (the supersymmetric algebra of the Z_2 - graded vector space L) onto the Z - graded superalgebra G(L) .

<u>Theorem</u> 1 (Poincaré, Birkhoff, Witt) [21 , 1]

The canonical homomorphism $\tilde{\varphi} : \tilde{U}(L) \longrightarrow G(L)$ is an isomorphism of Z - graded superalgebras.

As will be explained at the end of this section this important theorem can be proved by generalizing the arguments used in the Lie algebra case [9]. Let us, however, first draw some conclusions. Due to the known structure of the algebra $\tilde{U}(L)$ (see $n^o 2$) the theorem may be reformulated as follows :

<u>Corollary</u> 1

Let L be a Lie superalgebra and let σ be the canonical mapping of L into the enveloping algebra $U(L)$ of L. Suppose we are given a basis $(E_i)_{i\in I}$ of L such that all elements E_i are homogeneous and such that the index set I is totally ordered.

If $(i_1,\ldots,i_r)$ runs through all finite sequences in I such that

$$r \geqslant 0 \text{ arbitrary} \tag{2.37,a}$$

$$i_1 \leqslant i_2 \leqslant \ldots\ldots \leqslant i_r \tag{2.37,b}$$

$$i_p < i_{p+1} \text{ if } E_{i_p} \text{ and } E_{i_{p+1}} \text{ are odd ,} \tag{2.37,c}$$

then the products

$$\sigma(E_{i_1})\,\sigma(E_{i_2})\,\ldots\ldots\,\sigma(E_{i_r}) \tag{2.38}$$

form a basis of the vector space $U(L)$. (For $r = 0$ the product (2.38) is by definition equal to 1.)

Corollary 2

The canonical mapping $\sigma : L \longrightarrow U(L)$ is injective.

Due to this corollary we are able to identify L with a graded subspace of $U(L)$ by means of the mapping σ.

Corollary 3

Let L be a Lie superalgebra, let L' be a graded subalgebra of L and let $g : L' \longrightarrow L$ be the injection. Furthermore, let σ (resp. σ') be the canonical mapping of L into $U(L)$ (resp. of L' into $U(L')$).

According to corollary 2 of proposition 1 there exists a canonical homomorphism of superalgebras

$$\bar{g} : U(L') \longrightarrow U(L) \tag{2.39,a}$$

such that

$$\sigma \circ g = \bar{g} \circ \sigma' \quad , \quad \bar{g}(1) = 1 \ . \tag{2.39,b}$$

This homomorphism is injective.

By means of the homomorphism $\bar{g}$ the algebra $U(L)$ may be considered as a graded left or right $U(L')$ - module. Both of these modules are free (i.e.

both modules have a basis).

More precisely, let $(E_j)_{j\in J}$ be a family of homogeneous elements of L such that the images of the elements E_j under the canonical mapping $L \longrightarrow L/L'$ form a basis of the vector space L/L'. We assume that the index set J is totally ordered.

If $(j_1,\ldots,j_r)$ runs through all finite sequences in J such that

$$r \geqslant 0 \text{ arbitrary} \tag{2.40,a}$$

$$j_1 \leqslant j_2 \leqslant \ldots\ldots \leqslant j_r \tag{2.40,b}$$

$$j_p < j_{p+1} \text{ if } E_{j_p} \text{ and } E_{j_{p+1}} \text{ are odd ,} \tag{2.40,c}$$

then the products

$$\sigma(E_{j_1})\,\sigma(E_{j_2})\,\ldots\ldots\,\sigma(E_{j_r}) \tag{2.41}$$

form a basis of the left as well as of the right $U(L')$ - module $U(L)$.

The canonical homomorphism $\bar{g}$ enables us to identify the enveloping algebra $U(L')$ of L' with a graded subalgebra of $U(L)$.

Note that corollary 3 may be applied in particular to the special case $L' = L_{\bar{0}}$. Then $U(L_{\bar{0}})$ is the usual enveloping algebra of the Lie algebra $L_{\bar{0}}$ and $(E_j)_{j\in J}$ is a basis of the vector space $L_{\bar{1}}$. (One should be careful not to confuse $U(L_{\bar{0}})$ with the even subalgebra $U(L)_{\bar{0}}$ of $U(L)$.)

Finally, let us comment on the proof of the Poincaré, Birkhoff, Witt theorem. As stated above it is sufficient to prove corollary 1.

For every integer $r \geqslant 0$ let H_r be the set of all finite sequences $(i_1,\ldots,i_r)$ in I such that (2.37) is satisfied (in particular, $H_0 = \{\phi\}$) ; we define $H = \bigcup_{r\geqslant 0} H_r$.

Let $N = (i_1,\ldots,i_r)$ be an element of H and let $i \in I$. We write $i \leqslant N$ if and only if $(i,i_1,\ldots,i_r)$ is an element of H. In particular, we have $i \leqslant \phi$ for all $i \in I$.

Next we consider the tensor product

$$P = S(L_{\bar{0}}) \otimes \Lambda L_{\bar{1}} \tag{2.42}$$

which has been discussed in $n^o 2$. Recall that $P = \bigoplus_{n \in Z} P_n$ is a Z-graded superalgebra with $P_n = \{0\}$ if $n \leqslant -1$. For all $n \in Z$ we introduce

$$P^n = \bigoplus_{m \leqslant n} P_m \quad . \tag{2.43}$$

Define for every $i \in I$

$$F_i = \begin{cases} E_i \otimes 1 & \text{if } E_i \in L_{\bar{0}} \\ 1 \otimes E_i & \text{if } E_i \in L_{\bar{1}} \end{cases} \tag{2.44}$$

and, furthermore, for every $r \geqslant 0$ and every element $N = (i_1, \ldots, i_r) \in H_r$

$$F_N = F_{i_1} \cdot \ldots \cdot F_{i_r} \tag{2.45}$$

(by convention, $F_\phi = 1$). If N runs through all sequences from H_r the elements F_N form a basis of P_r.

The following lemma is the decisive step in proving the corollary 1.

Lemma 1

Denote the degree of the element E_i, $i \in I$, by $d(i)$. For every positive integer $r \geqslant 0$ there exists a unique linear mapping

$$f_r : L \otimes P^r \longrightarrow P \tag{2.46}$$

such that for all $i, j \in I$

$$f_r(E_i \otimes F_N) = F_i F_N \qquad \text{if } 0 \leqslant s \leqslant r \text{ and } N \in H_s,\ i \leqslant N \tag{2.47}$$

$$f_r(E_i \otimes F_N) - F_i F_N \in P^s \quad \text{if } 0 \leqslant s \leqslant r \text{ and } N \in H_s \tag{2.48}$$

$$\begin{aligned} & f_r(E_i \otimes f_r(E_j \otimes F_N)) \\ = {} & (-1)^{d(i)\,d(j)} f_r(E_j \otimes f_r(E_i \otimes F_N)) + f_r(\langle E_i, E_j \rangle \otimes F_N) \\ & \qquad \text{if } 0 \leqslant s \leqslant r-1 \text{ and } N \in H_s \ . \end{aligned} \tag{2.49}$$

The mapping f_r is homogeneous of degree zero (with respect to the Z_2-gradations). If $r \geqslant 1$ the restriction of f_r onto $L \otimes P^{r-1}$ is equal to f_{r-1}.

The lemma can be proved by induction with respect to r, we shall not go into the details.

From lemma 1 we deduce

Lemma 2

There exists a graded representation g of the Lie superalgebra L in the Z_2-graded vector space P such that for every $i \in I$

$$g(E_i)\, F_N = F_i\, F_N \qquad \text{if } N \in H \text{ and } i \leqslant N \tag{2.50}$$

$$g(E_i)\, F_N - F_i\, F_N \in P^r \quad \text{if } r \geqslant 0 \text{ and } N \in H_r \ . \tag{2.51}$$

Proof

Let $(f_r)_{r \geqslant 0}$ be the family of linear mappings defined in lemma 1. There exists a unique even linear mapping

$$f : L \otimes P \longrightarrow P \tag{2.52}$$

such that for every integer $r \geqslant 0$ the restriction of f onto $L \otimes P^r$ is equal to f_r. Define the mapping $g : L \longrightarrow \mathrm{Hom}(P)$ by

$$g(A)\, Z = f(A \otimes Z) \quad \text{for all } A \in L\, ,\ Z \in P\ . \tag{2.53}$$

Then g meets our requirements.

The corollary 1 now follows immediately. It is evident that the products $\sigma(E_{i_1}) \ldots . \sigma(E_{i_r})$ with $(i_1, \ldots, i_r) \in H$ generate the vector space $U(L)$. Let

$$\bar{g} : U(L) \longrightarrow \mathrm{Hom}(P) \tag{2.54}$$

be the homomorphism of superalgebras such that

$$g = \bar{g} \circ \sigma \quad , \quad \bar{g}(1) = \mathrm{id}_P \ . \tag{2.55}$$

If $N = (i_1, \ldots, i_r) \in H$ then

$$\bar{g}(\sigma(E_{i_1}) \ldots . \sigma(E_{i_r}))\, F_\phi = F_N \ . \tag{2.56}$$

Since the F_N, $N \in H$, are linearly independent, the same holds true for the products $\sigma(E_{i_1}) \ldots . \sigma(E_{i_r})$ with $(i_1, \ldots, i_r) \in H$.

4. The enveloping algebra as a Hopf superalgebra

Let U(L) be the enveloping algebra of a Lie superalgebra L. In this section we shall see that U(L) has a natural coalgebra structure [19]; combined with the algebra structure this converts U(L) into a Hopf superalgebra [19, 10]. In the present work only the most elementary results on this subject will be needed. For a more thorough discussion the reader is referred to [16].

Let $\sigma : L \longrightarrow U(L)$ be the canonical mapping. Evidently, the diagonal mapping

$$L \longrightarrow L \times L \quad , \quad A \longrightarrow (A, A) \quad \text{if } A \in L \tag{2.57}$$

is a homomorphism of Lie superalgebras. Hence, due to the corollary 2 of proposition 1 as well as to proposition 2 there exists a unique homomorphism of superalgebras

$$c : U(L) \longrightarrow U(L) \mathbin{\overline{\otimes}} U(L) \tag{2.58,a}$$

such that

$$c(\sigma(A)) = \sigma(A) \otimes 1 + 1 \otimes \sigma(A) \quad \text{for all } A \in L \tag{2.58,b}$$

$$c(1) = 1 \otimes 1 \ . \tag{2.58,c}$$

The homomorphism c is called the *coproduct* of the enveloping algebra U(L). It is easy to verify the following properties.

a) c is coassociative.

This is to say that

$$(c \otimes id_U) \circ c = (id_U \otimes c) \circ c \ , \tag{2.59}$$

both sides being algebra-homomorphisms of U(L) into $U(L) \mathbin{\overline{\otimes}} U(L) \mathbin{\overline{\otimes}} U(L)$. (Recall that the algebras $(U(L) \mathbin{\overline{\otimes}} U(L)) \mathbin{\overline{\otimes}} U(L)$, $U(L) \mathbin{\overline{\otimes}} (U(L) \mathbin{\overline{\otimes}} U(L))$ and $U(L) \mathbin{\overline{\otimes}} U(L) \mathbin{\overline{\otimes}} U(L)$ are canonically isomorphic.)

b) c is super-cocommutative.

Let s be the unique automorphism of the superalgebra $U(L) \mathbin{\overline{\otimes}} U(L)$ such that

$$s(X \otimes Y) = (-1)^{\xi\eta} Y \otimes X \tag{2.60}$$

for all $X \in U(L)_\xi$, $Y \in U(L)_\eta$; $\xi , \eta \in Z_2$.

Then our statement means that

$$s \circ c = c . \tag{2.61}$$

c) There exists a counit.

Consider K as an associative superalgebra (the odd subspace of K being equal to {0}). There exists a unique homomorphism of superalgebras

$$\varepsilon : U(L) \longrightarrow K \tag{2.62,a}$$

such that

$$\varepsilon \circ \sigma = 0 \quad , \quad \varepsilon(1) = 1 . \tag{2.62,b}$$

Identifying $U(L) \otimes K$ and $K \otimes U(L)$ with $U(L)$ canonically we have

$$(\varepsilon \otimes id_U) \circ c = (id_U \otimes \varepsilon) \circ c = id_U . \tag{2.63}$$

The homomorphism ε is called the *counit* of $U(L)$.

In view of the properties a) and c) the superalgebra $U(L)$, equipped with the coproduct c , is what may be called a *Hopf superalgebra.*

d) Let

$$\mu : U(L) \bar{\otimes} U(L) \longrightarrow U(L) \tag{2.64}$$

be the linear mapping defined by the multiplication in $U(L)$ and let

$$\bar{\varepsilon} : U(L) \longrightarrow U(L) \tag{2.65,a}$$

be defined by

$$\bar{\varepsilon}(X) = \varepsilon(X) \cdot 1_U \quad \text{for all } X \in U(L) . \tag{2.65,b}$$

If $\theta : U(L) \longrightarrow U(L)$ is the mapping introduced in (2.11) then

$$\mu \circ (\theta \otimes id_U) \circ c = \mu \circ (id_U \otimes \theta) \circ c = \bar{\varepsilon} . \tag{2.66}$$

A linear mapping θ with this property is called an *antipode.*

e) c is compatible with the filtration $(U^n(L))_{n\in Z}$ of U(L) (see n^o3). This means

$$c(U^n(L)) \subset \sum_{r+s=n} U^r(L) \otimes U^s(L) \qquad (2.67)$$

for all n.

f) Let L' be a second Lie superalgebra and let c' be its coproduct. If $g : L \longrightarrow L'$ is a homomorphism of Lie superalgebras and if $\bar{g} : U(L) \longrightarrow U(L')$ is the corresponding homomorphism of associative superalgebras, then

$$c' \circ \bar{g} = (\bar{g} \otimes \bar{g}) \circ c \ , \qquad (2.68)$$

i.e. $\bar{g}$ is a homomorphism of coalgebras, too.

§ 3 Representations of Lie superalgebras

In this paragraph we shall describe the standard constructions which are usually carried out with representations of Lie superalgebras. The discussion of induced and produced representations (which might also be included here) will be deferred to the next paragraph.

As might be anticipated the contents of this paragraph are merely a transcription of results which are well-known in the Lie algebra case; in fact all that we have to do is to add the appropriate sign factors. Nevertheless, for later reference it is helpful to have all this collected at some place.

1. The connection between representations of L and U(L)

In the definition 2 of §1 we have introduced the notion of a graded representation of a Lie superalgebra. Let us repeat this definition in a language which is adjusted to the results of §2 .

Definition 1

Let L be a Lie superalgebra and let V be a Z_2 - graded vector space. Recall that Hom(V) has a natural Z_2 - gradation which converts it into an associative superalgebra.
A *graded representation* ρ of L in V is an even linear mapping

$$\rho : L \longrightarrow \mathrm{Hom}(V) \qquad (3.1,a)$$

such that

$$\rho(\langle A , B \rangle) = \rho(A)\,\rho(B) - (-1)^{\alpha\beta}\,\rho(B)\,\rho(A) \qquad (3.1,b)$$

$$\text{for all } A \in L_\alpha ,\ B \in L_\beta ;\ \alpha , \beta \in Z_2 .$$

A Z_2 - graded vector space equipped with a graded representation of L is called a (left) *graded* L - *module*.

Remark 1)

The definition 1 makes sense even if the vector space V is not graded provided we drop the requirement that ρ should be even. In the present work we shall have no occasion to discuss these "non-graded" representations.

Let $U(L)$ be the enveloping algebra of L and let $\sigma : L \longrightarrow U(L)$ be the canonical mapping. From now on we shall identify L with a graded subspace of $U(L)$ by means of the mapping σ (see the remark below corollary 2 to theorem 1 in §2). Under this identification σ is just the injection of L into $U(L)$.

Let ρ be a graded representation of L in some graded vector space V . Due to the universal property of $U(L)$ there exists a unique homomorphism of associative superalgebras

$$\bar{\rho} : U(L) \longrightarrow \mathrm{Hom}(V) \tag{3.2,a}$$

which extends ρ , i.e. such that

$$\bar{\rho}(A) = \rho(A) \quad \text{for all } A \in L \quad , \quad \bar{\rho}(1) = \mathrm{id} \quad . \tag{3.2,b}$$

In particular, we have

$$\bar{\rho}(U(L)_\alpha) V_\beta \subset V_{\alpha+\beta} \quad \text{for all } \alpha , \beta \in Z_2 \quad . \tag{3.3}$$

Hence $\bar{\rho}$ is a graded representation of the associative superalgebra $U(L)$ in the graded vector space V or, using still another language, V is a Z_2 - graded left $U(L)$ - module.

Conversely, suppose we are given some Z_2 - graded left $U(L)$ - module V ; let

$$\omega : U(L) \longrightarrow \mathrm{Hom}(V) \tag{3.4}$$

be the corresponding homomorphism of associative superalgebras (according to our conventions we have $\omega(1) = \mathrm{id}$). Then the restriction ρ of ω to L is a graded representation of L in V and $\bar{\rho} = \omega$.

In view of this discussion the concepts of a "graded representation of L ", a "graded L - module" and a "left graded $U(L)$ - module" are completely equivalent. It depends on the circumstances which language is preferred (to make the formulation more suggestive).

Having described the definitions let us now simplify our notation. First of all we shall drop the bar and write ρ instead of $\bar{\rho}$. Normally it will be obvious from the context which representation of L in a vector space V is going to be considered. In this case we shall write X_V instead of $\rho(X)$, for all $X \in U(L)$. If this linear mapping acts on some element $y \in V$ we shall frequently even drop the subscript V. Hence in the following all three of the notations

$$\rho(X)\,y \;=\; X_V\,y \;=\; X\,y \tag{3.5}$$

will be used depending on which degree of precision might be necessary.

Example 1)

The vector space K has a trivial structure of a graded L-module: The Z_2-gradation of K is defined by

$$K_{\bar{0}} = K \quad , \quad K_{\bar{1}} = \{0\} \tag{3.6}$$

and the representation of L in K is equal to zero. If ε is the counit of U(L) (see §2, $n^0 4$) then

$$X_K\,a \;=\; \varepsilon(X)\,a \quad \text{for all } X \in U(L)\,,\ a \in K\ . \tag{3.7}$$

According to our general conventions on graded algebraic structures a *homomorphism* of a graded L-module V into a graded L-module W is an *even* linear mapping

$$f : V \longrightarrow W \tag{3.8,a}$$

such that

$$f(A\,y) \;=\; A\,f(y) \quad \text{for all } A \in L\,,\ y \in V\ . \tag{3.8,b}$$

It follows that

$$f(X\,y) \;=\; X\,f(y) \quad \text{for all } X \in U(L)\,,\ y \in V\ , \tag{3.8,c}$$

hence that f is a homomorphism of graded U(L)-modules. It is now obvious how to define an isomorphism of graded L-modules.

Remark 2)

Let $V = V_{\bar{0}} \oplus V_{\bar{1}}$ be a Z_2 - graded vector space and let $V' = V'_{\bar{0}} \oplus V'_{\bar{1}}$ be the Z_2 - graded vector space whose underlying vector space is equal to that of V but whose gradation is defined by

$$V'_{\bar{0}} = V_{\bar{1}} \quad , \quad V'_{\bar{1}} = V_{\bar{0}} \; . \tag{3.9}$$

It is obvious that the superalgebras Hom(V) and Hom(V') coincide. Consequently, a graded representation of L in V is also a graded representation of L in V', and vice versa. Nevertheless, these two representations are not necessarily isomorphic.

The concepts of a graded submodule, a graded quotient module, the direct sum, the simplicity (i.e. irreducibility), the semi-simplicity (i.e. complete reducibility) ..... of graded L - modules are self-explanatory and need not be defined separately.

2. Canonical constructions with L - modules

A. Extension of the base field

Let L be a Lie superalgebra and let V be a graded L - module. Furthermore, let E be a field containing K. Recall that $E \underset{K}{\otimes} L$ is a Lie superalgebra over E. The vector space $E \underset{K}{\otimes} V$ over E has a natural structure of a graded $E \underset{K}{\otimes} L$ - module: The Z_2 - gradation of $E \underset{K}{\otimes} V$ is defined by

$$(E \underset{K}{\otimes} V)_\alpha = E \underset{K}{\otimes} V_\alpha \quad \text{if } \alpha \in Z_2 \tag{3.10}$$

and the $E \underset{K}{\otimes} L$ - module structure is given by

$$(a \otimes A)(b \otimes x) = ab \otimes Ax \tag{3.11}$$

$$\text{for all } a, b \in E; \; A \in L; \; x \in V \; .$$

Remark 3)

According to n^o1 we may consider $E \underset{K}{\otimes} V$ as a graded $U(E \underset{K}{\otimes} L)$ - module.

On the other hand, V is a graded U(L) - module; hence, by extending the base field, $E \otimes_K V$ has a natural structure of an $E \otimes_K U(L)$ - module. These two module structures on $E \otimes_K V$ correspond to each other under the canonical isomorphism of $U(E \otimes_K L)$ onto $E \otimes_K U(L)$.

For later reference we mention the following proposition which is easily derived from standard results in representation theory [22]:

Proposition 1

Using the notation introduced above we assume that V is finite-dimensional. Then the graded L - module V is completely reducible if and only if the graded $E \otimes_K L$ - module $E \otimes_K V$ is completely reducible.

B. The tensor product of graded L - modules

Let L and L' be two Lie superalgebras. Suppose we are given a graded L - module V and a graded L'- module V'. Considering V as a graded U(L) - module and V' as a graded U(L') - module we know (see chapter 0, §2, 8)) that $V \otimes V'$ has a natural structure of a graded $U(L) \bar{\otimes} U(L')$ - module. In fact, the Z_2 - gradation is given by

$$(V \otimes V')_\alpha = \bigoplus_{\beta+\gamma=\alpha} (V_\beta \otimes V'_\gamma) \quad \text{if } \alpha \in Z_2 \tag{3.12}$$

and the module structure is defined by

$$(X \otimes X')(y \otimes y') = (-1)^{\xi'\eta} (Xy) \otimes (X'y') \tag{3.13}$$

for all $X \in U(L)$, $X' \in U(L')_{\xi'}$, $y \in V_\eta$, $y' \in V'$; $\xi', \eta \in Z_2$.

The graded tensor product $U(L) \bar{\otimes} U(L')$ is canonically isomorphic to the enveloping algebra of $L \times L'$ (see §2, proposition 2), hence $V \otimes V'$ is equipped with a natural graded $L \times L'$- module structure:

$$(A, A')(y \otimes y') = (Ay) \otimes y' + (-1)^{\alpha'\eta} y \otimes (A'y') \tag{3.14}$$

for all $A \in L$, $A' \in L'_{\alpha'}$, $y \in V_\eta$, $y' \in V'$; $\alpha', \eta \in Z_2$.

This graded $L \times L'$ - module $V \otimes V'$ is called the *tensor product of the graded* L *- module* V *with the graded* L' *- module* V'.

Suppose now that $L = L'$. Using the diagonal homomorphism $A \longrightarrow (A, A)$ of L into $L \times L$ we deduce from (3.14) the following graded L - module structure on $V \otimes V'$

$$A(y \otimes y') = (Ay) \otimes y' + (-1)^{\alpha\eta} y \otimes (Ay') \tag{3.15}$$

$$\text{for all } A \in L_\alpha ,\ y \in V_\eta ,\ y' \in V' ;\ \alpha , \eta \in Z_2 .$$

The graded L - module which has been obtained is called the *tensor product of the graded* L *- module* V *with the graded* L *- module* V'. (This name is not consistent with the notation introduced below (3.14), however, it is easy to avoid a confusion.)

Remark 4)

The $U(L)$ - module structure of $V \otimes V'$ corresponding to (3.15) is obtained from the $U(L) \bar{\otimes} U(L)$ - module structure of $V \otimes V'$ by restriction of the algebra of scalars via the coproduct of $U(L)$ (see §2, n°4).

We observe that in equation (3.15) the two factors V and V' are treated differently. Hence the canonical isomorphism of $V \otimes V'$ onto $V' \otimes V$ has to be modified. It is easy to check that the linear mapping

$$s : V \otimes V' \longrightarrow V' \otimes V \tag{3.16,a}$$

which is defined by

$$s(y \otimes y') = (-1)^{\eta\eta'} y' \otimes y \tag{3.16,b}$$

$$\text{for all } y \in V_\eta ,\ y' \in V'_{\eta'} ;\ \eta , \eta' \in Z_2$$

is an isomorphism of graded L - modules.

The constructions described above are easily generalized to the case of more than two factors. We shall not carry this out here but rather give an application of this generalization.

Let V be a Z_2 - graded vector space. Recall (see §2, n°2) that the tensor algebra $T(V)$ of V has a natural structure of a Z-graded superalgebra. Evidently, the general linear Lie superalgebra $pl(V)$ of V (see §1, example 3)) has a natural graded representation in V. The aforementioned generalization of (3.15) then yields a graded representation of $pl(V)$ in $T_n(V)$, for every integer $n \geqslant 1$. Considering $T_0(V)$ as a trivial graded $pl(V)$-module (see n°1, example 1)) and forming the direct sum of all these representations we obtain a graded representation $H \longrightarrow D_H^T$, $H \in pl(V)$, of $pl(V)$ in $T(V)$. It is easy to see that D_H^T is, for every $H \in pl(V)$, a superderivation (see §1, example 4)) of the superalgebra $T(V)$.

Next we consider the special case in which V is a Lie superalgebra L. Then $A \longrightarrow D_{\mathrm{ad}\,A}^T$, $A \in L$, is a graded representation of L in $T(L)$, called canonical. On the other hand, let $\mathfrak{D}(L)$ denote the Lie superalgebra of superderivations of L, let $H \in \mathfrak{D}(L)$ and let D_H^T be the corresponding superderivation of $T(L)$. If J is the two-sided ideal of $T(L)$ introduced at the beginning of §2, n°1, then $D_H^T(J) \subset J$. Hence, by going to the quotient, the superderivation D_H^T of $T(L)$ defines a superderivation D_H^U of the enveloping algebra $U(L)$. It is obvious that $H \longrightarrow D_H^U$, $H \in \mathfrak{D}(L)$, is a graded representation of $\mathfrak{D}(L)$ in $U(L)$. In particular, $A \longrightarrow D_{\mathrm{ad}\,A}^U$, $A \in L$, is a graded representation of L in $U(L)$ which is also called canonical. According to our conventions we shall write A_U instead of $D_{\mathrm{ad}\,A}^U$. Then

$$A_U(Z) = AZ - (-1)^{\alpha\zeta} ZA \qquad (3.17)$$

$$\text{for all } A \in L_\alpha,\ Z \in U(L)_\zeta;\ \alpha, \zeta \in Z_2,$$

i.e. A_U is the inner superderivation of $U(L)$ defined by the element A of L. Note that by definition the canonical mapping $T(L) \longrightarrow U(L)$ is a homomorphism of graded L-modules.

The graded L-module structure on $U(L)$ which we have constructed above may also be obtained as follows. Let $\theta : U(L) \longrightarrow U(L)$ be the linear mapping introduced in (2.11). Then there exists a unique structure of a graded $U(L) \mathbin{\overline{\otimes}} U(L)$-module on $U(L)$ such that

$$(X \otimes Y)_U(Z) = (-1)^{\eta\zeta} XZ(\theta Y) \qquad (3.18)$$

$$\text{for all } X \in U(L),\ Y \in U(L)_\eta,\ Z \in U(L)_\zeta;\ \eta, \zeta \in Z_2.$$

Using the canonical isomorphism of $U(L \times L)$ onto $U(L) \bar{\otimes} U(L)$ as well as the diagonal injection $L \longrightarrow L \times L$ we are again led to (3.17).

C. Representations in spaces of multilinear mappings

Let L and L' be two Lie superalgebras. Suppose we are given a graded L-module V and a graded L'-module V'. We consider V as a graded U(L)-module and V' as a graded U(L')-module. Then Hom(V,V'), the space of all K-linear mappings of V into V', has a natural structure of a graded $U(L) \bar{\otimes} U(L')$-module. In fact, the Z_2-gradation is chosen to be

$$\mathrm{Hom}(V,V')_\alpha = \{ h \in \mathrm{Hom}(V,V') \mid h(V_\beta) \subset V'_{\alpha+\beta} , \beta \in Z_2 \} \tag{3.19}$$

for all $\alpha \in Z_2$, and the module structure is defined by

$$(X \otimes X')g = (-1)^{\xi(\xi'+\gamma)} X'_{V'} \circ g \circ (\theta X)_V \tag{3.20}$$

for all $X \in U(L)_\xi$, $X' \in U(L')_{\xi'}$, $g \in \mathrm{Hom}(V,V')_\gamma$; $\xi, \xi', \gamma \in Z_2$.

Here again θ is the linear mapping introduced in (2.11). The corresponding representation of $L \times L'$ is given by

$$(A, A')g = A'_{V'} \circ g - (-1)^{\alpha\gamma} g \circ A_V \tag{3.21}$$

for all $A \in L_\alpha$, $A' \in L'$, $g \in \mathrm{Hom}(V,V')_\gamma$; $\alpha, \gamma \in Z_2$.

Once again the case L = L' is particularly important. In this case we derive from (3.21) the following structure of a graded L-module on Hom(V,V') :

$$Ag = A_{V'} \circ g - (-1)^{\alpha\gamma} g \circ A_V \tag{3.22}$$

for all $A \in L_\alpha$, $g \in \mathrm{Hom}(V,V')_\gamma$; $\alpha, \gamma \in Z_2$.

Now let V' = K be the trivial L-module (see $n^o 1$, example 1)). Then Hom(V,K) = V^* is the dual vector space of V, its Z_2-gradation is given by

$$(V^*)_\alpha = \{ g \in V^* \mid g(V_{\alpha+\bar{1}}) = \{0\} \} \tag{3.23}$$

and (3.22) leads to the following action of L on V^*

$$A g = -(-1)^{\alpha\gamma} g \circ A_V \tag{3.24}$$

for all $A \in L_\alpha$, $g \in (V^*)_\gamma$; $\alpha, \gamma \in Z_2$.

We call this L - module V* the L - module *contragredient* to V . Note that in this case

$$X g = (-1)^{\xi\gamma} g \circ (\theta X)_V \tag{3.25}$$

for all $X \in U(L)_\xi$, $g \in (V^*)_\gamma$; $\xi, \gamma \in Z_2$.

The results of the present and of the preceding section may now be combined to construct several other representations. As an example we consider a space of bilinear mappings. Suppose then that V , V', W are three graded L - modules. The vector space of bilinear mappings from $V \times V'$ into W will be denoted by B(V,V';W) . By definition of $V \otimes V'$ this space is canonically isomorphic to $\mathrm{Hom}(V \otimes V', W)$. Therefore, the natural gradation of B(V,V';W) is given by

$$B(V,V';W)_\beta = \{ b \in B(V,V';W) \mid b(V_\xi, V'_{\xi'}) \subset W_{\beta+\xi+\xi'} ;\ \xi, \xi' \in Z_2 \} \tag{3.26}$$

for all $\beta \in Z_2$, and the graded representation of L in B(V,V';W) is defined by

$$(A b)(x,x') = A\, b(x,x') - (-1)^{\alpha\beta} b(Ax,x') - (-1)^{\alpha(\beta+\xi)} b(x,Ax') \tag{3.27}$$

for all $A \in L_\alpha$, $b \in B(V,V';W)_\beta$, $x \in V_\xi$, $x' \in V'$; $\alpha, \beta, \xi \in Z_2$.

As is well-known there is a number of canonical homomorphisms between the tensor products of vector spaces and the spaces of multilinear mappings. Starting from graded L - modules we have to discuss whether these mappings are homomorphisms of graded L - modules.

Let $V, W, V_1, \ldots, V_n$ be graded L - modules. All the tensor products and spaces of multilinear mappings are equipped with a structure of a graded L - module according to the aforementioned constructions. Then the following linear mappings are homomorphisms of graded L - modules.

$$\eta : V \longrightarrow V^{**}$$

$$(\eta(x))(g) = (-1)^{\xi\gamma} g(x) \tag{3.28}$$

$$\text{if } x \in V_\xi \text{ , } g \in (V^*)_\gamma \text{ ; } \xi \text{ , } \gamma \in Z_2$$

$$\lambda : B(V,W;K) \longrightarrow \mathrm{Hom}(V,W^*)$$

$$((\lambda(b))(x))(y) = b(x,y) \tag{3.29}$$

$$\text{if } b \in B(V,W;K) \text{ , } x \in V \text{ , } y \in W$$

$$\mu : \mathrm{Hom}(V,W^*) \longrightarrow (V \otimes W)^*$$

$$(\mu(h))(x \otimes y) = (h(x))(y) \tag{3.30}$$

$$\text{if } h \in \mathrm{Hom}(V,W^*) \text{ , } x \in V \text{ , } y \in W$$

$$\tau : W \otimes V^* \longrightarrow \mathrm{Hom}(V,W)$$

$$(\tau(y \otimes g))(x) = y\, g(x) \tag{3.31}$$

$$\text{if } y \in W \text{ , } g \in V^* \text{ , } x \in V$$

$$\omega : V_n^* \otimes \ldots\ldots \otimes V_1^* \longrightarrow B(V_1,\ldots\ldots,V_n;K)$$

$$(\omega(g_n \otimes \ldots\ldots \otimes g_1))(x_1,\ldots\ldots,x_n) = \prod_{i=1}^{n} g_i(x_i) \tag{3.32}$$

$$\text{if } g_i \in V_i^* \text{ , } x_i \in V_i \text{ ; } 1 \leqslant i \leqslant n \text{ .}$$

Here $B(V_1,\ldots\ldots,V_n;K)$ is the space of n - linear forms on $V_1 \times \ldots\ldots \times V_n$. We draw the reader's attention to the sign factor in (3.28) as well as to the order in which the L - modules have been written down.

Finally, let L' be a second Lie superalgebra and let V' and W' be two graded L' - modules. Recall (see chapter 0, §2, 3)) that for all

$$g \in \mathrm{Hom}(V,W)_\gamma \quad , \quad g' \in \mathrm{Hom}(V',W')_{\gamma'} \quad ; \quad \gamma \text{ , } \gamma' \in Z_2 \tag{3.33,a}$$

we have defined a linear mapping

$$g \mathbin{\bar{\otimes}} g' : V \otimes V' \longrightarrow W \otimes W' \tag{3.33,b}$$

by the condition that

$$(g \mathbin{\bar{\otimes}} g')(x \otimes x') = (-1)^{\gamma'\xi}\, g(x) \otimes g'(x') \tag{3.33,c}$$

$$\text{for all } x \in V_\xi\,,\ x' \in V'\,;\ \xi \in Z_2\,.$$

We extend this definition bilinearly (in g, g') to the general case where g and g' are not necessarily homogeneous. Then the linear mapping

$$\psi : \mathrm{Hom}(V,W) \otimes \mathrm{Hom}(V',W') \longrightarrow \mathrm{Hom}(V \otimes V', W \otimes W') \tag{3.34,a}$$

which satisfies

$$\psi(g \otimes g') = g \mathbin{\bar{\otimes}} g' \tag{3.34,b}$$

$$\text{for all } g \in \mathrm{Hom}(V,W)\ ,\ g' \in \mathrm{Hom}(V',W')$$

is a homomorphism of graded $L \times L'$ - modules.

Remark 5)

Equation (3.24) suggests the following definition.

Definition 2

Let V be a Z_2 - graded vector space. There exists a unique even linear mapping

$$\mathrm{Hom}(V) \longrightarrow \mathrm{Hom}(V^*)\ ,\quad \text{written as}\quad A \longrightarrow {}^{T}\!A \tag{3.35,a}$$

such that

$${}^{T}\!A(g) = (-1)^{\alpha\gamma}\, g \circ A \tag{3.35,b}$$

$$\text{for all } A \in \mathrm{Hom}(V)_\alpha\,,\ g \in (V^*)_\gamma\,;\ \alpha, \gamma \in Z_2\,.$$

If $A \in \mathrm{Hom}(V)$, then ${}^{T}\!A$ is called the *supertranspose of* A . (The usual transpose of A will be denoted by ${}^{t}\!A$.)

We know from (3.24) that $A \longrightarrow -{}^{T}\!A$ is a homomorphism of $pl(V)$ into $pl(V^*)$; in fact, it is easy to see that, more generally,

$$^{T}(AB) = (-1)^{\alpha\beta}\, {}^{T}B\, {}^{T}A \tag{3.36}$$

for all $A \in \mathrm{Hom}(V)_\alpha$, $B \in \mathrm{Hom}(V)_\beta$; $\alpha, \beta \in Z_2$.

Since the mapping η in (3.28) is a homomorphism of L - modules we conclude that

$$^{TT}A \circ \eta = \eta \circ A \quad \text{for all } A \in \mathrm{Hom}(V) . \tag{3.37}$$

If V is finite-dimensional and if V** is identified with V by means of the *usual* canonical isomorphism equation (3.37) may be rewritten as

$$^{TT}A = (-1)^\alpha A \quad \text{for all } A \in \mathrm{Hom}(V)_\alpha , \alpha \in Z_2 . \tag{3.38}$$

3. Invariants

Definition 3

Let L be a Lie superalgebra and let V be a graded L - module. An element $x \in V$ is called *invariant* with respect to the given representation of L in V (or simply L - invariant) if

$$A_V(x) = 0 \quad \text{for all } A \in L . \tag{3.39}$$

The set of all L - invariant elements of V is sometimes denoted by V^L. An element of V is L - invariant if and only if its homogeneous components are L - invariant. Hence V^L is a graded subspace of V .

Examples

2) Let L be a Lie superalgebra and let V be a graded L - module. If x is a *homogeneous* element of V then

$$H = \{ A \in L \mid A_V(x) = 0 \} \tag{3.40}$$

is a graded subalgebra of L and x is H - invariant. We call H the graded subalgebra of L consisting of those elements of L which leave x invariant.

3) Let V and W be two graded L - modules. An element $g \in \mathrm{Hom}(V,W)_\gamma$, $\gamma \in Z_2$,

is L - invariant if and only if

$$X_W \circ g = (-1)^{\xi\gamma} g \circ X_V \qquad (3.41)$$
$$\text{for all } X \in U(L)_\xi, \ \xi \in Z_2 .$$

The graded vector space of all L - invariant linear mappings of V into W will be denoted by $\mathrm{Hom}_L(V,W)$. In the case where the language of U(L) - modules is more natural we shall also write $\mathrm{Hom}_{U(L)}(V,W)$ instead of $\mathrm{Hom}_L(V,W)$ and call the elements of this space U(L) - invariant. Note that the elements of $\mathrm{Hom}_L(V,W)_{\bar{0}}$ are the homomorphisms of the graded L - module V into the graded L - module W .

Now let L' be a second Lie superalgebra and let V' and W' be two graded L' - modules. If $g \in \mathrm{Hom}_L(V,W)_\gamma$ and $g' \in \mathrm{Hom}_{L'}(V',W')_{\gamma'}$ with $\gamma, \gamma' \in Z_2$, then $g \bar{\otimes} g'$ (see (3.33)) is an element of $\mathrm{Hom}_{L \times L'}(V \otimes V', W \otimes W')_{\gamma+\gamma'}$.

In the graded situation *Schur's lemma* takes the following form.

Proposition 2

Suppose that the field K is algebraically closed. Let L be a Lie superalgebra and let V be a finite-dimensional simple graded L - module. Then

$$\mathrm{Hom}_L(V)_{\bar{0}} = K \cdot \mathrm{id} \ , \quad \mathrm{Hom}_L(V)_{\bar{1}} = K \cdot u \ , \qquad (3.42)$$

where either $u = 0$ or else $u^2 = -\mathrm{id}$.

4) Suppose we are given three graded L - modules V , W and U . A bilinear mapping $b : V \times W \longrightarrow U$ which is homogeneous of degree β is L - invariant if and only if

$$A\,b(x,y) = (-1)^{\alpha\beta} b(Ax,y) + (-1)^{\alpha(\beta+\xi)} b(x,Ay) \qquad (3.43)$$
$$\text{for all } A \in L_\alpha, \ x \in V_\xi, \ y \in W ; \ \alpha, \xi \in Z_2 .$$

For every bilinear mapping $b : V \times W \longrightarrow U$ we define a bilinear mapping $sb : W \times V \longrightarrow U$ by

$$sb(y,x) = (-1)^{\xi\eta} b(x,y) \qquad (3.44)$$
$$\text{for all } x \in V_\xi, \ y \in W_\eta ; \ \xi, \eta \in Z_2 .$$

The mapping $b \longrightarrow sb$ of $B(V,W;U)$ into $B(W,V;U)$ is an isomorphism of graded L-modules (see (3.16)), hence b is L-invariant if and only if sb is L-invariant. This leads to the following definition.

Definition 4

Let U and V be two Z_2-graded vector spaces. A bilinear mapping b of $V \times V$ into U is called *supersymmetric / skew-supersymmetric* if $sb = \pm b$, i.e. if

$$b(y,x) = \pm(-1)^{\xi\eta} b(x,y) \tag{3.45}$$
$$\text{for all } x \in V_\xi ,\ y \in V_\eta ;\ \xi, \eta \in Z_2 .$$

For example, recall that the product mapping in a Lie superalgebra is skew-supersymmetric.

Next we shall mention some special cases.

a) Let T be any superalgebra. We choose $V = W = U = T$ and $L = \mathfrak{D}(T)$ (the Lie superalgebra of superderivations of T; see §1, example 4)). Then the product mapping $T \times T \longrightarrow T$ is even and (by definition) $\mathfrak{D}(L)$-invariant. The same holds true for the product mappings with more than two factors.

b) Let $U = K$ be the trivial L-module (see n°1, example 1)). A bilinear form b on $V \times W$ is L-invariant if and only if

$$b(Ax,y) + (-1)^{\alpha\xi} b(x,Ay) = 0 \tag{3.46}$$
$$\text{for all } A \in L_\alpha ,\ x \in V_\xi ,\ y \in W ;\ \alpha, \xi \in Z_2 .$$

Let $\tilde{b} : V \longrightarrow W^*$ be the linear mapping which is associated canonically with b, i.e.

$$(\tilde{b}(x))(y) = b(x,y) \quad \text{for all } x \in V ,\ y \in W \tag{3.47}$$

(see (3.29)). Then b is L-invariant if and only if $\tilde{b}$ is L-invariant.

c) In example b) we choose $V = W = L$ (equipped with the adjoint representation). Then the condition (3.46) is equivalent to

$$b(\langle A,B\rangle, C) = b(A, \langle B,C\rangle) \quad \text{for all } A, B, C \in L . \tag{3.48}$$

We shall now introduce an important class of invariant n - linear forms on a Lie superalgebra. Let V be a finite-dimensional Z_2 - graded vector space and let

$$\gamma : V \longrightarrow V \tag{3.49,a}$$

be the linear mapping which satisfies

$$\gamma(x) = (-1)^{\xi} x \quad \text{if } x \in V_{\xi} \; ; \; \xi \in Z_2 \; . \tag{3.49,b}$$

We define a linear form str on the general linear Lie superalgebra pl(V) by

$$\mathrm{str}(A) = \mathrm{Tr}(\gamma A) \quad \text{for all } A \in \mathrm{pl}(V) \tag{3.50}$$

and call str the *supertrace*. The linear form str is even and pl(V) - invariant :

$$\mathrm{str}(\langle A , B \rangle) = 0 \quad \text{for all } A , B \in \mathrm{pl}(V) \tag{3.51}$$

or, equivalently,

$$\mathrm{str}(AB) = (-1)^{\alpha\beta} \mathrm{str}(BA) \tag{3.52}$$

$$\text{for all } A \in \mathrm{Hom}(V)_{\alpha} , \; B \in \mathrm{Hom}(V)_{\beta} \; ; \; \alpha , \beta \in Z_2 \; .$$

We conclude :

Proposition 3

Let L be a Lie superalgebra and let V be a finite-dimensional graded L - module. The n - linear form on L which is defined by

$$(A^1, \ldots\ldots, A^n) \longrightarrow \mathrm{str}(A^1_V \ldots\ldots A^n_V) \tag{3.53}$$

$$\text{for all } A^i \in L \; , \; 1 \leqslant i \leqslant n \; ,$$

is even and L - invariant (with respect to the adjoint representation of L).

The n - linear form defined in proposition 3 is called the *n - linear form associated with the graded L - module* V . The most important case is obtained if we choose V = L and n = 2 .

Definition 5

Let L be a finite-dimensional Lie superalgebra and let γ be the automorphism of L such that

$$\gamma(A) = (-1)^{\alpha} A \quad \text{for all } A \in L_{\alpha} \, ; \, \alpha \in Z_2 \, . \tag{3.54}$$

The bilinear form ϕ on L which is defined by

$$\phi(A,B) = \mathrm{str}(\mathrm{ad}\,A\,\mathrm{ad}\,B) = \mathrm{Tr}(\gamma\,\mathrm{ad}\,A\,\mathrm{ad}\,B) \tag{3.55}$$

for all $A, B \in L$ is called the *Killing form of* L ; it is even, invariant and supersymmetric.

In the next proposition we shall construct some important L - invariant elements of the enveloping algebra U(L) . Recall that U(L) has a natural structure of a graded L - module which has been defined in section B (see (3.17)).

Proposition 4

Let L be a Lie superalgebra and let H be a finite-dimensional graded ideal of L . Suppose we are given a non-degenerate homogeneous bilinear form b on H of degree β and a homogeneous n - linear form h on H of degree ζ . We assume that b and h are L - invariant if the ideal H is considered as a graded submodule of L .

Let $(E_i)_{1 \leqslant i \leqslant p}$ be a basis of H consisting of homogeneous elements and let ε_i be the degree of E_i . We introduce a second basis $(F_j)_{1 \leqslant j \leqslant p}$ of H by the condition that

$$b(F_j, E_i) = \delta_{ij} \quad ; \quad 1 \leqslant i,j \leqslant p \, . \tag{3.56}$$

Define

$$X = \sum_{1 \leqslant i_1, \ldots, i_n \leqslant p} (-1)^{\beta\sigma(i_1,\ldots,i_n)} h(E_{i_1}, \ldots, E_{i_n}) F_{i_n} \cdots F_{i_1} \tag{3.57}$$

$$\text{with } \sigma(i_1,\ldots,i_n) = \sum_{s=1}^{n} s\,\varepsilon_{i_s} \, .$$

Then X is a homogeneous L - invariant element of U(L) of degree $\zeta + n\beta$, hence

$$XY = (-1)^{(\zeta + n\beta)\eta}\, YX \quad \text{for all } Y \in U(L)_\eta \; ; \; \eta \in Z_2 \; . \qquad (3.58)$$

The element X does not depend on the choice of the basis $(E_i)_{1 \leq i \leq p}$.

Proof

Use the canonical homomorphisms (3.32) , (3.47) and $T(L) \longrightarrow U(L)$.

The element X in (3.57) is called a *generalized Casimir element* of U(L), it "supercommutes" with every element of U(L) .

§ 4 INDUCED AND PRODUCED REPRESENTATIONS

In the present paragraph we shall generalize part of Blattner's classical work on induced and produced representations of Lie algebras [23] to the graded situation (see also [24]). With the formal apparatus as developed in §§ 2, 3 at hand this amounts to simply "replacing the non-graded objects by the corresponding graded ones".

Our main objective is to prove a realization theorem for transitive filtered Lie superalgebras which generalizes the Lie algebra analogue by Guillemin and Sternberg [25]. This theorem will be an important tool for the classification of simple Lie superalgebras.

Next we would like to draw the reader's attention to the following peculiarity of the graded case. Let L be a Lie superalgebra and let V be a $L_{\bar{0}}$ - module. If L and V are finite-dimensional then the same holds true for the corresponding induced and produced L - modules. This provides us with a multitude of finite-dimensional graded L - modules and in particular yields a proof of the Ado theorem for Lie superalgebras.

Before going into the details let us introduce some notation which will be used throughout this paragraph. We suppose that we are given a Lie superalgebra L and a graded subalgebra L' of L. Let U(L) (resp. U(L')) be the enveloping algebra of L (resp. L'). According to §2 (see the corollaries 2 and 3 to theorem 1) we shall assume that L, L' and U(L') are canonically embedded in U(L). Then U(L) has an obvious structure of a left (resp. right) graded U(L') - module which is defined by left (resp. right) multiplication. Moreover, the Poincaré, Birkhoff, Witt theorem implies the following proposition (which is in fact nothing but a reformulation of the corollary 3 to theorem 1 in §2).

Proposition 1

Let $(E_j)_{j \in J}$ be a family of homogeneous elements of L such that the images of the elements E_j under the canonical mapping $L \longrightarrow L/L'$ form a basis of the vector space L/L'. We assume that the index set J is totally ordered.

Let H be the set of all finite sequences $(j_1, \ldots, j_r)$ in J such that

$$r \geq 0 \text{ arbitrary} \tag{4.1,a}$$

$$j_1 \leq j_2 \leq \cdots\cdots \leq j_r \tag{4.1,b}$$

$$j_p < j_{p+1} \text{ if } E_{j_p} \text{ and } E_{j_{p+1}} \text{ are odd .} \tag{4.1,c}$$

If $N = (j_1, \ldots, j_r)$ is a sequence of this type we define

$$E_N = E_{j_1} E_{j_2} \cdots\cdots E_{j_r} \tag{4.2}$$

(by convention $E_\phi = 1$).

Then the family $(E_N)_{N \in H}$ is a basis of the left as well as of the right $U(L')$ - module $U(L)$.

1. Induced representations

Let V be a left graded $U(L')$ - module. Regarding $U(L)$ as a right graded $U(L')$ - module we can construct the tensor product

$$\overline{V} = U(L) \underset{U(L')}{\otimes} V . \tag{4.3}$$

It is known [18] that $\overline{V}$ has a natural structure of a Z_2 - graded vector space, the subspace $\overline{V}_\gamma$, $\gamma \in Z_2$, being spanned by the tensors of the form $X \otimes y$ with $X \in U(L)_\xi$, $y \in V_\eta$; $\xi, \eta \in Z_2$, $\xi + \eta = \gamma$. Now it is obvious that $\overline{V}$ has a structure of a left graded $U(L)$ - module which satisfies

$$X(Y \otimes y) = (XY) \otimes y \tag{4.4}$$

$$\text{for all } X, Y \in U(L) , \; y \in V .$$

This graded $U(L)$ - module $\overline{V}$ is called the *graded* $U(L)$ - *module induced from the graded* $U(L')$ - *module* V .

Let us define an even linear mapping

$$\alpha : V \longrightarrow \overline{V} \tag{4.5,a}$$

by

$$\alpha(y) = 1 \otimes y \quad \text{if } y \in V . \tag{4.5,b}$$

Obviously, α is $U(L')$ - invariant (see §3, n°3, example 3)):

$$\alpha(X'y) = X'\alpha(y) \quad \text{for all } X' \in U(L')\ ,\ y \in V\ . \tag{4.6}$$

The pair $(\bar{V},\alpha)$ may be characterized by the following universal property.

Proposition 2

Let $g : V \longrightarrow W$ be a $U(L')$ - invariant linear mapping of V into any left graded $U(L)$ - module W. Then there exists a unique $U(L)$ - invariant linear mapping $\bar{g} : \bar{V} \longrightarrow W$ such that

$$g = \bar{g} \circ \alpha\ . \tag{4.7}$$

The proof is straightforward. Let us remark that if g is homogeneous of degree γ, $\gamma \in Z_2$, then the same holds true for $\bar{g}$ and we have

$$\bar{g}(X \otimes y) = (-1)^{\gamma\xi} X\, g(y) \tag{4.8}$$

$$\text{for all } X \in U(L)_\xi\ ,\ y \in V\ ;\ \xi \in Z_2\ .$$

Next we shall exploit proposition 1. Using standard results on tensor products we conclude :

Lemma 1

We use the notation introduced in proposition 1. For every sequence $N \in H$ the K - linear mapping

$$V \longrightarrow \bar{V} \quad , \quad y \longrightarrow E_N\,\alpha(y) = E_N \otimes y \tag{4.9}$$

is injective; in particular, α itself is injective. The vector space $\bar{V}$ is the direct sum of its subspaces $E_N\,\alpha(V)$, $N \in H$.

This lemma may now be applied to give a proof of the Ado theorem for Lie superalgebras. In fact, let $L' = L_{\bar{0}}$ and let V be any $L_{\bar{0}}$ - module. We introduce a trivial Z_2 - gradation in V by defining

$$V_{\bar{0}} = V \quad , \quad V_{\bar{1}} = \{0\}\ . \tag{4.10}$$

Then the induced L - module $\bar{V}$ is well-defined and lemma 1 implies :

a) If the $L_{\bar{0}}$ - module V is faithful, then the L - module $\bar{V}$ is faithful, too, provided that $V \neq \{0\}$.

b) If the vector spaces $L_{\bar{1}}$ and V are finite-dimensional, then the same holds true for $\bar{V}$.

Using the Ado theorem for Lie algebras we conclude [21] :

Theorem 1 (Ado)

Every finite-dimensional Lie superalgebra has a faithful finite-dimensional graded representation.

2. Produced representations

Let V be a left graded $U(L')$ - module. Regarding $U(L)$ as a left graded $U(L')$ - module we consider the Z_2 - graded vector space

$$\hat{V} = \mathrm{Hom}_{U(L')}(U(L), V) . \tag{4.11}$$

Recall (see §3, n°3, example 3)) that $\hat{V}$ is the graded subspace of $\mathrm{Hom}_K(U(L),V)$ consisting of those K - linear mappings $h : U(L) \longrightarrow V$ which are $U(L')$ - invariant; if h is homogeneous of degree η this is to say that

$$X' h(X) = (-1)^{\xi'\eta} h(X'X) \tag{4.12}$$

$$\text{for all } X' \in U(L')_{\xi'} , \; X \in U(L) ; \; \xi' \in Z_2 .$$

Now it is easy to see that on the Z_2 - graded vector space $\mathrm{Hom}_K(U(L),V)$ there exists a unique structure of a left graded $U(L)$ - module such that

$$(Xh)(Z) = (-1)^{\xi(\eta+\zeta)} h(ZX) \tag{4.13}$$

$$\text{for all } X \in U(L)_{\xi} , \; h \in \mathrm{Hom}_K(U(L),V)_{\eta} , \; Z \in U(L)_{\zeta} ; \; \xi, \eta, \zeta \in Z_2 .$$

Moreover, the graded subspace $\hat{V}$ of $\mathrm{Hom}_K(U(L),V)$ is in fact a graded $U(L)$ - submodule of this $U(L)$ - module $\mathrm{Hom}_K(U(L),V)$.

Thus we have defined on $\hat{V}$ the structure of a left graded $U(L)$ - module; this graded $U(L)$ - module $\hat{V}$ is called the *graded* $U(L)$ *- module produced from the graded* $U(L')$ *- module* V .

Remark 1)

By restriction of the algebra of scalars from $U(L)$ to $U(L')$ we may consider $\hat{V}$ as a left graded $U(L')$ - module. If in the following we shall speak of the $U(L')$ - module $\hat{V}$ we shall always refer to this structure (and not to the one which has been used in §3, n°3 to define the space $\mathrm{Hom}_{U(L')}(U(L),V)$; in fact, this latter $U(L')$ - module structure on $\hat{V}$ is trivial by the very definition of $\hat{V}$).

Let us define an even linear mapping

$$\beta : \hat{V} \longrightarrow V \tag{4.14,a}$$

by

$$\beta(h) = h(1) \quad \text{if } h \in \hat{V} . \tag{4.14,b}$$

This mapping is $U(L')$ - invariant : If $X' \in U(L')_{\xi'}$, $h \in \hat{V}_\eta$; $\xi', \eta \in Z_2$, we have

$$\beta(X'h) = (X'h)(1) = (-1)^{\xi'\eta} h(X') = X'h(1) = X'\beta(h) . \tag{4.15}$$

The pair $(\hat{V},\beta)$ may be characterized by the following universal property.

Proposition 3

Let W be any left graded $U(L)$ - module and let $g : W \longrightarrow V$ be a $U(L')$ - invariant linear mapping. Then there exists a unique $U(L)$ - invariant linear mapping $\hat{g} : W \longrightarrow \hat{V}$ such that

$$g = \beta \circ \hat{g} . \tag{4.16}$$

Again the proof is straightforward. We remark that if g is homogeneous of degree γ , $\gamma \in Z_2$, then the same holds true for $\hat{g}$; furthermore, we have

$$(\hat{g}(y))(X) = (-1)^{\xi\eta} g(Xy) \tag{4.17}$$

$$\text{for all } y \in W_\eta , \; X \in U(L)_\xi \; ; \; \eta , \xi \in Z_2 .$$

The concepts of induced and produced modules are dual to each other. In fact, recall the definition in (2.11) of the linear mapping θ of $U(L)$

into itself; furthermore, recall that we have defined in §3, n°3, C the notion of a contragredient L-module (in particular, see (3.25)). It is now an easy task to prove the following proposition.

Proposition 4

Let V be a left graded U(L')-module and let V* be the left graded U(L')-module contragredient to it. The linear mapping

$$\left(U(L) \underset{U(L')}{\otimes} V \right)^* \longrightarrow \mathrm{Hom}_{U(L')}(U(L),V^*) \tag{4.18}$$

which associates with every linear form $\ell \in \left(U(L) \underset{U(L')}{\otimes} V \right)^*$ the linear mapping

$$\ell' : U(L) \longrightarrow V^* \tag{4.19,a}$$

defined by

$$(\ell'(X))(y) = \ell((\theta X) \otimes y) \quad \text{for all } X \in U(L)\,,\ y \in V \tag{4.19,b}$$

is an isomorphism of the graded U(L)-module $\left(U(L) \underset{U(L')}{\otimes} V \right)^*$ (contragredient to the U(L)-module induced from V) onto the graded U(L)-module $\mathrm{Hom}_{U(L')}(U(L),V^*)$ (which is produced from the contragredient U(L')-module V*).

3. Additional structures on produced modules : Filtration and multiplication

The filtration and coalgebra structure of the enveloping algebra U(L) lead to some additional structures on produced modules which we are now going to describe.

As before, let V be a left graded U(L')-module and let $\hat{V}$ be the left graded U(L)-module produced from it, $\hat{V} = \mathrm{Hom}_{U(L')}(U(L),V)$. In §2, n°3 we have defined the canonical filtration $(U^n(L))_{n\in Z}$ of the enveloping algebra U(L). This immediately yields a (downward) filtration of $\hat{V}$, as follows. We define for all $n \in Z$

$$\hat{V}^n = \{ h \in \hat{V} \mid h(U^{n-1}(L)) = \{0\} \} \,. \tag{4.20}$$

Obviously, the following statements hold true.

a) The $\hat{V}^n$ are Z_2 - graded subspaces of $\hat{V}$.

b) $$\hat{V}^n \supset \hat{V}^m \quad \text{if } n \leqslant m \tag{4.21}$$

c) $$\hat{V}^n = \hat{V} \quad \text{if } n \leqslant 0 \tag{4.22}$$

d) $$\bigcap_{n \geqslant 0} \hat{V}^n = \{0\} \tag{4.23}$$

e) $$U^m(L)\,\hat{V}^n \subset \hat{V}^{n-m} \quad \text{for all } n, m \in Z . \tag{4.24}$$

The family $(\hat{V}^n)_{n \in Z}$ is called the *canonical filtration* of the produced module $\hat{V}$.

Next we shall exploit the coalgebra structure of U(L) (see §2, n°4) to introduce a multiplicative structure on produced modules.

The general procedure is well-known [19]. Let V, V' and W be three left graded U(L') - modules and let

$$p : V \times V' \longrightarrow W \tag{4.25}$$

be a L' - invariant bilinear mapping. The linear mapping $V \otimes V' \longrightarrow W$ defined by p will be denoted by $\bar{p}$. Of course, $\bar{p}$ is U(L') - invariant.

Now let

$$g \in \mathrm{Hom}_{U(L')}(U(L), V) \quad , \quad g' \in \mathrm{Hom}_{U(L')}(U(L), V') . \tag{4.26}$$

In (3.33) we have defined a linear mapping

$$g \mathbin{\bar{\otimes}} g' : U(L) \mathbin{\bar{\otimes}} U(L) \longrightarrow V \otimes V' . \tag{4.27}$$

Since g and g' are U(L') - invariant we know that in fact

$$g \mathbin{\bar{\otimes}} g' \in \mathrm{Hom}_{U(L') \bar{\otimes} U(L')}(U(L) \mathbin{\bar{\otimes}} U(L) , V \otimes V') \tag{4.28}$$

(see §3, n°3, example 3)).

If c is the coproduct of the enveloping algebra U(L) (see §2, n°4) we may now define a linear mapping

$$\hat{p}(g,g') : U(L) \longrightarrow W \tag{4.29,a}$$

by composing the linear mappings of the following sequence :

$$\hat{p}(g,g') : U(L) \xrightarrow{c} U(L) \bar{\otimes} U(L) \xrightarrow{g \bar{\otimes} g'} V \otimes V' \xrightarrow{\bar{p}} W . \tag{4.29,b}$$

From the remarks made above it is easy to see that this mapping is $U(L')$ - invariant, i.e.

$$\hat{p}(g,g') \in \mathrm{Hom}_{U(L')}(U(L),W) . \tag{4.30}$$

Hence we have defined a bilinear mapping

$$\hat{p} : \hat{V} \times \hat{V}' \longrightarrow \hat{W} . \tag{4.31}$$

It is now straightforward to prove the following properties of $\hat{p}$.

α) If p is homogeneous of degree π, $\pi \in Z_2$, then the same holds true for $\hat{p}$.

β) The bilinear mapping $\hat{p}$ is L - invariant.

γ) The mapping $\hat{p}$ is consistent with the canonical filtrations in the sense that

$$\hat{p}(\hat{V}^n, \hat{V}'^m) \subset \hat{W}^{n+m} \quad \text{for all } n, m \in Z . \tag{4.32}$$

The coassociativity and super-cocommutativity of the coproduct c imply certain general associativity and super-commutativity properties of our construction, however, we do not want to describe them here.

All the results which we have obtained thus far will now be applied to the following important special case. Let V = K be the trivial one-dimensional graded $U(L')$ - module (see §3, $n^0 1$, example 1)). We consider the produced $U(L)$ - module

$$F = \mathrm{Hom}_{U(L')}(U(L),K) . \tag{4.33}$$

Evidently, the product mapping

$$p : K \times K \longrightarrow K \tag{4.34}$$

is even and L' - invariant. Therefore, the corresponding bilinear mapping

$$\hat{p} : F \times F \longrightarrow F \tag{4.35}$$

defines on F a superalgebra structure.

Proposition 5

1) The superalgebra F is associative and super-commutative, furthermore, the counit ε of U(L) is the unit element of F .

2) With respect to the produced graded U(L) - module structure of F the elements of L act as superderivations of the superalgebra F .

3) The canonical filtration $(F^n)_{n \in Z}$ of F satisfies :

a) The F^n are Z_2 - graded subspaces of F .

b) $$F^n \supset F^m \quad \text{if } n \leqslant m \tag{4.36}$$

c) $$F^n = F \quad \text{if } n \leqslant 0 \tag{4.37}$$

d) $$\bigcap_{n \geqslant 0} F^n = \{0\} \tag{4.38}$$

e) $$U^m(L)\, F^n \subset F^{n-m} \quad \text{for all } n , m \in Z \tag{4.39}$$

f) $$F^n F^m \subset F^{n+m} \quad \text{for all } n , m \in Z \tag{4.40}$$

g) $$F = K \cdot \varepsilon \oplus F^1 \ . \tag{4.41}$$

Proof

1) As indicated above the coassociativity of the coproduct c implies the associativity of F . Similarly, the super-commutativity of F follows from the symmetry of p and the super-cocommutativity of c .

2) The fact that the elements of L act as superderivations of F is equivalent to the L - invariance of $\hat{p}$ (see the property β) above).

3) The statements a) - e) are valid for general produced L - modules, relation f) is nothing but a special case of equation (4.32), and relation g) is obvious.

4. Some non-canonical constructions

In this section we shall apply the Poincaré, Birkhoff, Witt theorem to get information on the filtered Z_2 - graded vector space underlying a produced U(L) - module.

Let $\tilde{U}(L/L')$ be the supersymmetric algebra of the Z_2 - graded vector space L/L' (see §2, n°2). Recall that $\tilde{U}(L/L')$ is nothing but the enveloping algebra U(L/L') where L/L' is given the abelian Lie superalgebra structure.

In the following we shall use the notation introduced in proposition 1. For every $j \in J$ we denote the canonical image of E_j in L/L' by $\tilde{E}_j$. By definition, $\tilde{E}_j$ is homogeneous of the same degree as E_j and the family $(\tilde{E}_j)_{j\in J}$ is a basis of the vector space L/L'. If $N = (j_1, \dots, j_r) \in H$ we define

$$\tilde{E}_N = \tilde{E}_{j_1} \tilde{E}_{j_2} \cdot \ldots \cdot \tilde{E}_{j_r} . \tag{4.42}$$

Then $(\tilde{E}_N)_{N\in H}$ is a basis of the vector space $\tilde{U}(L/L')$. Hence there exists a unique K - linear mapping

$$\tau : \tilde{U}(L/L') \longrightarrow U(L) \tag{4.43,a}$$

such that

$$\tau(\tilde{E}_N) = E_N \quad \text{for all } N \in H . \tag{4.43,b}$$

Lemma 2

The mapping τ is even, injective and a homomorphism of coalgebras. More explicitly, let c (resp. $\tilde{c}$) be the coproduct and let ε (resp. $\tilde{\varepsilon}$) be the counit of U(L) (resp. $\tilde{U}(L/L')$). Then we have

$$(\tau \otimes \tau) \circ \tilde{c} = c \circ \tau \tag{4.44}$$

$$\tilde{\varepsilon} = \varepsilon \circ \tau . \tag{4.45}$$

The proof is straightforward.

Now let V be any left graded U(L') - module. We may regard $\mathrm{Hom}_K(\tilde{U}(L/L'),V)$ as the left graded $\tilde{U}(L/L')$ - module produced from the graded {0} - module V . Hence our general constructions with produced modules may be applied to $\mathrm{Hom}_K(\tilde{U}(L/L'),V)$. In particular, $\mathrm{Hom}_K(\tilde{U}(L/L'),V)$ has a canonical filtration.

Lemma 3

The linear mapping

$$\pi : \mathrm{Hom}_{U(L')}(U(L),V) \longrightarrow \mathrm{Hom}_K(\tilde{U}(L/L'),V) \tag{4.46,a}$$

defined by

$$\pi(h) = h \circ \tau \quad \text{if } h \in \mathrm{Hom}_{U(L')}(U(L),V) \tag{4.46,b}$$

is an isomorphism of filtered Z_2 - graded vector spaces.

Once more the proof is straightforward; however, let us stress that it makes essential use of the Poincaré, Birkhoff, Witt theorem.

We are mainly interested in the special case in which V = K is the trivial graded U(L') - module (see §3, $n^{o}1$, example 1)). Considering

$$\tilde{F} = \mathrm{Hom}_K(\tilde{U}(L/L'),K) \tag{4.47}$$

as the left graded $\tilde{U}(L/L')$ - module produced from the graded {0} - module K we know that $\tilde{F}$ has a natural structure of a filtered associative superalgebra (see proposition 5). Then lemma 2 and lemma 3 combine to yield the following proposition.

Proposition 6

The linear mapping

$$\pi : F \longrightarrow \tilde{F} \tag{4.48,a}$$

defined by

$$\pi(h) = h \circ \tau \quad \text{if } h \in F \tag{4.48,b}$$

is an isomorphism of filtered superalgebras.

Note that the isomorphism π depends on the choice of the family $(E_j)_{j \in J}$, hence π is not canonical.

5. The Guillemin, Sternberg realization theorem

Let $F = \mathrm{Hom}_{U(L')}(U(L),K)$ be the filtered associative superalgebra which has been defined in section 3 and whose basic properties have been collected in proposition 5 . Let $\mathcal{D}(F)$ be the Lie superalgebra of superderivations of F (see §1, example 4)). We define for every $n \in Z$

$$D^n = \{ d \in \mathcal{D}(F) \mid d(F^r) \subset F^{n+r} \text{ for all } r \in Z \} . \tag{4.49}$$

Evidently, D^n is a Z_2 - graded subspace of $\mathcal{D}(F)$ and

$$D^n \supset D^m \quad \text{if } n \leqslant m \tag{4.50}$$

$$\bigcap_{n\in Z} D^n = \{0\} \tag{4.51}$$

$$\langle D^n, D^m \rangle \subset D^{n+m} \quad \text{for all } n, m \in Z . \tag{4.52}$$

Setting

$$\bigcup_{n\in Z} D^n = D \tag{4.53}$$

we conclude that D is a graded subalgebra of $\mathcal{D}(F)$ and that $(D^n)_{n\in Z}$ is a (decreasing) filtration of D . Since every superderivation of F annihilates the unit element of F we conclude from (4.41) that

$$D^0 F \subset F^1 . \tag{4.54}$$

Let us agree that a subalgebra G of D is filtered by setting

$$G^n = G \cap D^n \quad \text{for all } n \in Z . \tag{4.55}$$

On the other hand we define a family $(L^n)_{n\in Z}$ of subspaces of L as follows (see also chapter II, §1, lemma 7). Set

$$L^n = L \quad \text{if } n \leqslant -1 \tag{4.56}$$

$$L^0 = L' \tag{4.57}$$

and define inductively

$$L^{n+1} = \{ A \in L^n \mid \langle A , L \rangle \subset L^n \} \quad \text{if } n \geqslant 0 . \tag{4.58}$$

Evidently, all L^n are Z_2 - graded subspaces of L and

$$L^n \supset L^m \quad \text{if } n \leqslant m \ . \tag{4.59}$$

Moreover, an easy induction argument shows that

$$\langle L^n, L^m \rangle \subset L^{n+m} \quad \text{for all } n, m \in Z \ . \tag{4.60}$$

Thirdly we remark that $\bigcap_{n \in Z} L^n = \{0\}$ if and only if L' does not contain any non-zero graded ideal of L. If this is the case L is a transitive filtered Lie superalgebra in the sense of chapter II, §1, definition 3.

Finally let us recall that F carries a structure of a produced graded U(L) - module. Let ρ be the corresponding graded representation of L in F. According to proposition 5 (see part 2) and relation (4.39)) we have

$$\rho(L) \subset D^{-1} \ . \tag{4.61}$$

<u>Theorem</u> 2 (Blattner)

Suppose that L' does not contain any non-zero graded ideal of L.

Then the graded representation ρ of L in F defines an isomorphism of the filtered Lie superalgebra L onto a filtered graded subalgebra of D.

Let $\rho' : L \longrightarrow D$ be a homomorphism of Lie superalgebras such that

$$\rho'(A) - \rho(A) \in D^0 \quad \text{for all } A \in L \ . \tag{4.62}$$

Then there exists a unique automorphism ψ of the filtered superalgebra F such that

$$\psi \circ \rho'(A) = \rho(A) \circ \psi \quad \text{for all } A \in L \ . \tag{4.63}$$

Proof

We have already mentioned that $\rho(L) \subset D^{-1}$. Next let $A \in L$; we show that $\rho(A) \in D^0$ if and only if $A \in L'$. Suppose first that $A \in L'_\alpha$, $\alpha \in Z_2$, and let $g \in F^m_\gamma$, $X \in U^{m-1}(L)_\xi$ where $m \in Z$; $\gamma, \xi \in Z_2$. Then we have

$$X A = (-1)^{\xi\alpha} A X + Y \tag{4.64}$$

with some element $Y \in U^{m-1}(L)$. It follows that

$$(\rho(A)g)(X) = A\,g(X) + (-1)^{\alpha(\gamma+\xi)} g(Y) = 0 , \tag{4.65}$$

hence $\rho(A)g \in F^m$ and, therefore,

$$\rho(L') \subset D^0 . \tag{4.66}$$

To prove the converse we remark that $\rho^{-1}(D^0)$ is a graded subspace of L. Suppose that $A \in L_\alpha$, $\alpha \in Z_2$, but $A \notin L'$. Let $\tilde{A}$ be the canonical image of A in L/L'. Using the notation introduced in section 4 we may assume that there is an index $i \in J$ such that $A = E_i$, hence $\tilde{A} = \tilde{E}_i$ and

$$\tau(\tilde{A}) = A . \tag{4.67}$$

Evidently, there exists a homogeneous element $h \in \mathrm{Hom}_K(\tilde{U}(L/L'),K)$ (of degree $\gamma \in Z_2$, say) such that

$$h(\tilde{A}) \neq 0 . \tag{4.68}$$

Set $g = \pi^{-1}(h)$; then $g \in F_\gamma$ and

$$(\rho(A)g)(1) = (-1)^{\alpha\gamma} h(\tilde{A}) \neq 0 . \tag{4.69}$$

Therefore $\rho(A)g \notin F^1$, which implies $\rho(A) \notin D^0$ (see (4.54)). Thus we have shown that

$$\rho^{-1}(D^0) \subset L' . \tag{4.70}$$

It follows immediately that ρ is injective. In fact, $A \in \rho^{-1}(0)$ implies $\rho(A)F^m = \{0\} \subset F^m$ for all $m \in Z$, thus $\rho(A) \in D^0$ and hence $A \in L'$. Consequently

$$\rho^{-1}(0) \subset L' . \tag{4.71}$$

But L' does not contain any non-zero graded ideal of L, hence $\rho^{-1}(0) = \{0\}$.

We next show by induction that

$$\rho(L^n) = \rho(L) \cap D^n \quad \text{for all } n \in Z . \tag{4.72}$$

This relation has already been proved for $n \leq 0$. Suppose that (4.72) is known for some $n \geqslant 0$; we have to show that

$$\rho(L^{n+1}) = \rho(L) \cap D^{n+1} . \tag{4.73}$$

Let $A \in L_{\alpha}^{n+1}$, $\alpha \in Z_2$; we show that $\rho(A) \in D^{n+1}$. By definition

$$A \in L^n \text{ and } \langle A , B \rangle \in L^n \text{ for all } B \in L \tag{4.74}$$

and hence

$$\rho(A) \in D^n \text{ and } \rho(\langle A , B\rangle) \in D^n \text{ for all } B \in L . \tag{4.75}$$

We have to prove that

$$\rho(A)\, F^m \subset F^{m+n+1} \quad \text{for all } m \in Z . \tag{4.76}$$

Of course, this relation is obvious if $m \leqslant -n-1$. As a secondary induction, assume that

$$\rho(A)\, F^{m-1} \subset F^{m+n} \tag{4.77}$$

for some $m \geqslant -n$. Let $g \in F_{\gamma}^{m}$, $X \in U^{m+n-1}(L)_{\xi}$, $B \in L_{\beta}$ with $\gamma , \xi , \beta \in Z_2$. Then

$$\begin{aligned} &(\rho(A)g)(XB) \\ &= (-1)^{\beta(\gamma+\xi)} (\rho(A)\rho(B)g)(X) - (-1)^{\beta(\gamma+\xi)} (\rho(\langle A , B\rangle)\, g)(X) . \end{aligned} \tag{4.78}$$

Note that $\rho(B)g \in F^{m-1}$; hence (4.77) and (4.75) imply that $\rho(A)\rho(B)g$ and $\rho(\langle A , B\rangle)\, g$ are elements of F^{m+n} . Consequently, the right hand side of (4.78) vanishes.

But every element of $U^{m+n}(L)$ is a linear combination of 1 and of products of the form XB . Since $\rho(A) \in D^0$ by (4.75), we conclude from (4.54) that $(\rho(A)g)(1) = 0$. Hence we have shown that

$$(\rho(A)g)(U^{m+n}(L)) = \{0\} \tag{4.79}$$

and this implies (4.76) .

Consequently $\rho(A) \in D^{n+1}$ and, therefore,

$$\rho(L^{n+1}) \subset D^{n+1} . \tag{4.80}$$

Conversely, assume that $A \in L$ and $\rho(A) \in D^{n+1}$; we have to show that $A \in L^{n+1}$. Let $B \in L$; since $\rho(B) \in D^{-1}$, we have

$$\rho(\langle A , B\rangle) = \langle \rho(A) , \rho(B)\rangle \in D^n . \qquad (4.81)$$

Of course $\rho(A) \in D^n$. But then our induction hypothesis implies

$$A \in L^n \text{ and } \langle A , B\rangle \in L^n \quad \text{for all } B \in L \qquad (4.82)$$

which is equivalent to $A \in L^{n+1}$. Thus

$$\rho^{-1}(D^{n+1}) \subset L^{n+1} \qquad (4.83)$$

and the first part of the theorem is proved.

To prove the second part let $\rho' : L \longrightarrow D$ be a homomorphism of the type described in the theorem. Evidently,

$$\rho'(L) \subset D^{-1} \quad , \quad \rho'(L') \subset D^0 . \qquad (4.84)$$

Let β be the canonical linear form on F which is defined by

$$\beta(g) = g(1) \quad \text{if } g \in F . \qquad (4.85)$$

Because of (4.54) and $\beta(F^1) = \{0\}$ we have

$$\beta \circ \rho(A) = \beta \circ \rho'(A) = 0 \quad \text{for all } A \in L' . \qquad (4.86)$$

Let F' be the graded U(L) - module whose underlying Z_2 - graded vector space is equal to that of F but whose U(L) - module structure is given by ρ' (canonically extended to U(L)). According to (4.86), $\beta : F' \longrightarrow K$ is a L' - invariant even linear mapping. Since $F = \mathrm{Hom}_{U(L')}(U(L),K)$ is the graded U(L) - module produced from K we conclude from proposition 3 that there exists a unique U(L) - invariant linear mapping $\psi : F' \longrightarrow F$ such that

$$\beta = \beta \circ \psi . \qquad (4.87)$$

We know that ψ is even; more precisely, recall that

$$(\psi(g))(X) = (-1)^{\gamma\xi} (\rho'(X)g)(1) \qquad (4.88)$$
$$\text{for all } g \in F_\gamma , X \in U(L)_\xi ; \gamma , \xi \in Z_2 .$$

The U(L) - invariance of ψ is equivalent to

$$\psi \circ \rho'(X) = \rho(X) \circ \psi \quad \text{for all } X \in U(L) . \tag{4.89}$$

We have already mentioned that $\rho'(L) \subset D^{-1}$. It follows that

$$\rho'(U^n(L))F^m \subset F^{m-n} \quad \text{for all } n, m \in Z . \tag{4.90}$$

Now let $g \in F^m_\gamma$ and $X \in U^{m-1}(L)_\xi$ where $m \in Z$; $\gamma, \xi \in Z_2$. Then (4.90) yields $\rho'(X)g \in F^1$ and hence (see (4.88))

$$(\psi(g))(X) = (-1)^{\gamma\xi}(\rho'(X)g)(1) = 0 . \tag{4.91}$$

We conclude that $\psi(g) \in F^m$ and, therefore,

$$\psi(F^m) \subset F^m \quad \text{for all } m \in Z . \tag{4.92}$$

Let us next show that ψ is bijective. In fact, if $B \in L$, the relation (4.62) implies

$$(\rho'(B) - \rho(B))F^m \subset F^m \quad \text{for all } m \in Z . \tag{4.93}$$

Using (4.89) and (4.92) this yields

$$(\rho(B) \circ \psi - \psi \circ \rho(B))F^m \subset F^m \quad \text{for all } m \in Z . \tag{4.94}$$

If $n \geqslant 1$ is any positive integer and if $B_1, \ldots, B_n \in L$, the relation (4.94) may be iterated to give

$$(\rho(B_1 \ldots B_n) \circ \psi - \psi \circ \rho(B_1 \ldots B_n))F^m \subset F^{m-n+1} \quad \text{for all } m \in Z , \tag{4.95}$$

which implies

$$(\rho(X) \circ \psi - \psi \circ \rho(X))F^m \subset F^{m-n+1} \tag{4.96}$$
$$\text{for all } X \in U^n(L) \text{ and all } n, m \in Z .$$

Now let $g \in F^m_\gamma$ and $X \in U^m(L)_\xi$ where $m \in Z$; $\gamma, \xi \in Z_2$. Taking into account the relations (4.96) and (4.87) as well as the equation $\beta(F^1) = \{0\}$ we obtain

$$(\psi(g))(X) = (-1)^{\gamma\xi}(\beta \circ \rho(X) \circ \psi)g = (-1)^{\gamma\xi}(\beta \circ \rho(X))g = g(X). \tag{4.97}$$

It follows that

$$\psi(g) - g \in F^{m+1} \quad \text{for all } g \in F^m \text{ and all } m \in Z \ . \tag{4.98}$$

Obviously, this implies that ψ is injective. On the other hand, standard techniques (filtration completeness of F) may now be applied to prove that

$$\psi(F^m) = F^m \quad \text{for all } m \in Z \ . \tag{4.99}$$

We now show that ψ is multiplicative. In the proof we shall use the notation introduced in section 3. In particular, recall that $\overline{p} : K \otimes K \longrightarrow K$ is the canonical mapping and that c is the coproduct in U(L). Moreover, let

$$\mu : F \otimes F \longrightarrow F \tag{4.100}$$

denote the linear mapping defined by the multiplication in F and let

$$\Delta : U(D) \longrightarrow U(D) \mathbin{\overline{\otimes}} U(D) \tag{4.101}$$

be the coproduct of the enveloping algebra of D.

Of course, F is a graded D-module in the obvious way; hence F is a graded U(D)-module and $F \otimes F$ is a graded $U(D) \mathbin{\overline{\otimes}} U(D)$-module. The fact that the elements of D are superderivations of the algebra F implies

$$\mu \circ \Delta(Q)_{F \otimes F} = Q_F \circ \mu \quad \text{for all } Q \in U(D) \ . \tag{4.102}$$

The mappings ρ and ρ' are homomorphisms of the Lie superalgebra L into the Lie superalgebra D. Their canonical extensions to homomorphisms of the superalgebra U(L) into the superalgebra U(D) will be denoted by $\overline{\rho}$ and $\overline{\rho}'$, respectively. Then $\overline{\rho} \otimes \overline{\rho}$ and $\overline{\rho}' \otimes \overline{\rho}'$ are homomorphisms of the superalgebra $U(L) \mathbin{\overline{\otimes}} U(L)$ into the superalgebra $U(D) \mathbin{\overline{\otimes}} U(D)$ and we have

$$\Delta \circ \overline{\rho} = (\overline{\rho} \otimes \overline{\rho}) \circ c \tag{4.103}$$

$$\Delta \circ \overline{\rho}' = (\overline{\rho}' \otimes \overline{\rho}') \circ c \qquad (4.103)'$$

(see §2, n°4, f)).

The U(L)-invariance of ψ (i.e. equation (4.89)) yields

$$(\psi \otimes \psi) \circ ((\bar{\rho}' \otimes \bar{\rho}')(Z))_{F \otimes F} = ((\bar{\rho} \otimes \bar{\rho})(Z))_{F \otimes F} \circ (\psi \otimes \psi) \tag{4.104}$$
$$\text{for all } Z \in U(L) \bar{\otimes} U(L) \ .$$

Next it is easy to see that

$$\beta \circ \mu = \bar{p} \circ (\beta \otimes \beta) \ . \tag{4.105}$$

In view of (4.87) this equation implies

$$\beta \circ \mu = \beta \circ \mu \circ (\psi \otimes \psi) \ . \tag{4.106}$$

Now let $X \in U(L)$. Then

$$\begin{aligned} \beta \circ \rho(X) \circ \psi \circ \mu &= \beta \circ \psi \circ \rho'(X) \circ \mu \\ &= \beta \circ \bar{\rho}'(X)_F \circ \mu \\ &= \beta \circ \mu \circ (\Delta \circ \bar{\rho}'(X))_{F \otimes F} \\ &= \beta \circ \mu \circ (\psi \otimes \psi) \circ ((\bar{\rho}' \otimes \bar{\rho}') \circ c(X))_{F \otimes F} \\ &= \beta \circ \mu \circ ((\bar{\rho} \otimes \bar{\rho}) \circ c(X))_{F \otimes F} \circ (\psi \otimes \psi) \\ &= \beta \circ \mu \circ (\Delta \circ \bar{\rho}(X))_{F \otimes F} \circ (\psi \otimes \psi) \\ &= \beta \circ \rho(X) \circ \mu \circ (\psi \otimes \psi) \ . \end{aligned} \tag{4.107}$$

Since

$$(\beta \circ \rho(X))(g) = (-1)^{\xi\gamma} g(X) \tag{4.108}$$
$$\text{for all } X \in U(L)_\xi \, , \ g \in F_\gamma \, ; \ \xi, \gamma \in Z_2$$

we conclude from (4.107) that

$$\psi \circ \mu = \mu \circ (\psi \otimes \psi) \ , \tag{4.109}$$

i.e. ψ is multiplicative.

Finally, let us prove the uniqueness of ψ . Let ψ' be a second automorphism of the filtered superalgebra F such that (4.63) (and hence also (4.89)) is fulfilled with ψ' in place of ψ . If we can show that

$$\beta = \beta \circ \psi' \tag{4.110}$$

the uniqueness statement in proposition 3 will imply that $\psi' = \psi$.

Evidently,

$$\psi'(\varepsilon) = \varepsilon \quad \text{and} \quad \psi'(F^1) \subset F^1 \tag{4.111}$$

(where ε is the unit element of F, i.e. the counit of $U(L)$).

Now let $g \in F$. Obviously, $g - \beta(g)\varepsilon \in F^1$, hence also $\psi'(g) - \beta(g)\varepsilon \in F^1$ and, therefore,

$$\beta\,\psi'(g) = \beta(g) \quad . \tag{4.112}$$

Thus equation (4.110) is valid and the theorem is proved.

We shall now use the results of section 4 to derive an important corollary. Recall that $\tilde{F} = \mathrm{Hom}_K(\tilde{U}(L/L'),K)$ and that we have constructed an isomorphism

$$\pi : F \longrightarrow \tilde{F} \tag{4.113}$$

of filtered superalgebras (see proposition 6).

Starting from $\tilde{F}$ let us define the filtered Lie superalgebra $\tilde{D} = \bigcup_{n\in Z} \tilde{D}^n$ in the same way as $D = \bigcup_{n\in Z} D^n$ has been obtained from F. The prescription $d \longrightarrow \pi \circ d \circ \pi^{-1}$ defines an isomorphism $D \longrightarrow \tilde{D}$ of filtered Lie superalgebras. It follows that the mapping

$$\sigma : L \longrightarrow \tilde{D} \tag{4.114,a}$$

defined by

$$\sigma(A) = \pi \circ \rho(A) \circ \pi^{-1} \quad \text{for all } A \in L \tag{4.114,b}$$

is an isomorphism of the filtered Lie superalgebra L onto a filtered graded subalgebra of $\tilde{D}$.

Now $\tilde{F}$ has a structure of a produced graded $U(L/L')$ - module; let $\tilde{\rho}$ be the corresponding graded representation of the (by definition) abelian Lie superalgebra L/L' in $\tilde{F}$. For every element $A \in L$ let $\tilde{A}$ denote the canonical image of A in L/L'.

Lemma 4

We have

$$\sigma(A) - \tilde{\rho}(\tilde{A}) \in \tilde{D}^0 \quad \text{for all } A \in L \quad . \tag{4.115}$$

Sketch of the proof

We use the notation introduced in section 4 . Let $m \geq 0$ be any positive integer and let $h \in \tilde{F}_{\gamma}^{m+1}$, $\gamma \in Z_2$; finally, let N be a sequence from H consisting of at most m elements. Then it is not difficult to show that

$$(\sigma(E_j)h)(\tilde{E}_N) = (\tilde{\rho}(\tilde{E}_j)h)(\tilde{E}_N) \quad \text{for all } j \in J . \tag{4.116}$$

This implies

$$\sigma(E_j) - \tilde{\rho}(\tilde{E}_j) \in \tilde{D}^0 \quad \text{for all } j \in J . \tag{4.117}$$

On the other hand, if $A \in L'$ we have $\rho(A) \in D^0$ and hence $\sigma(A) \in \tilde{D}^0$, whereas $\tilde{A} = 0$.

The following corollary is now obvious.

Corollary 1 (Guillemin, Sternberg)

Suppose that L' does not contain any non-zero graded ideal of L .

1) There exist homomorphisms η of the Lie superalgebra L into the Lie superalgebra $\tilde{D}$ such that

$$\eta(A) - \tilde{\rho}(\tilde{A}) \in \tilde{D}^0 \quad \text{for all } A \in L . \tag{4.118}$$

2) Any such homomorphism is in fact an isomorphism of the filtered Lie superalgebra L onto a filtered graded subalgebra of $\tilde{D}$.

3) Any two such homomorphisms are conjugate under a uniquely determined automorphism of the filtered superalgebra $\tilde{F}$.

The importance of this corollary lies in the fact that the superalgebra $\tilde{U}(L/L')$ is well-known (see §2, n°2). For example, suppose that $L_{\bar{0}} \subset L'$. Then the even subspace of L/L' is equal to $\{0\}$, hence $\tilde{U}(L/L')$ is just the exterior algebra of the vector space L/L' . It follows :

Corollary 2

Suppose that $L_{\bar{0}} \subset L'$ and that L' does not contain any non-zero graded ideal of L . Then L is finite-dimensional if and only if L/L' is finite-dimensional. (See also chapter II, §6, theorem 2.)

CHAPTER II SIMPLE LIE SUPERALGEBRAS

This chapter is devoted to a detailed study of simple Lie superalgebras; in particular, we shall prove the classification theorem for all finite-dimensional simple Lie superalgebras over an algebraically closed field. The proof to be presented here requires a good knowledge of the transitive irreducible consistently Z-graded Lie superalgebras (see §1) of "depth one", hence the classification of these algebras is also included.

Convention

From §2 on we shall always assume that the *Lie superalgebras are finite-dimensional.*

§ 1 MISCELLANIES ON Z-GRADED AND FILTERED LIE SUPERALGEBRAS

In this paragraph we shall derive some elementary properties of Z-graded and filtered Lie superalgebras [3]. It is not our aim to present a "coherent theory" of these algebras here; rather we want to prepare the ground for the proofs of some deeper results which will be established in §§ 5-8.

1. Some definitions concerning Z-graded Lie superalgebras and a criterion for two bitransitive Lie superalgebras to be isomorphic

The following definition is a reformulation for the case of Lie superalgebras of the general definition 2 in chapter 0, §2

Definition 1

A Lie superalgebra G is called Z-graded if we are given a family $(G_n)_{n\in Z}$ of Z_2-graded subspaces of G such that

$$G = \bigoplus_{n\in Z} G_n \tag{1.1}$$

$$\langle G_n , G_m \rangle \subset G_{n+m} \quad \text{for all } n , m \in Z . \tag{1.2}$$

The gradation $(G_n)_{n\in Z}$ is called *consistent* with the Z_2-gradation of G (and the algebra G is called consistently Z - graded) if

$$G_{\bar{0}} = \bigoplus_{n\in Z} G_{2n} \quad ; \quad G_{\bar{1}} = \bigoplus_{n\in Z} G_{2n+1} \quad . \tag{1.3}$$

Let J be a subset of Z . If we shall speak of the Z - graded Lie superalgebra $G = \bigoplus_{n\in J} G_n$ this is to imply that G is a Z - graded Lie superalgebra such that $G_n = \{0\}$ if $n \notin J$.

From (1.2) we deduce that $\langle G_0 , G_n \rangle \subset G_n$ for all $n \in Z$; hence G_0 is a graded subalgebra of G and the adjoint representation of G induces a natural graded representation of G_0 in G_n , for all $n \in Z$.

Definition 2

Let $G = \bigoplus_{n\in Z} G_n$ be a Z - graded Lie superalgebra. The gradation $(G_n)_{n\in Z}$ (and the algebra G itself) will be called

a) *irreducible* if the graded representation of G_0 in G_{-1} is irreducible (in particular, $G_{-1} \neq \{0\}$),

b) *transitive* if

$$\{ A \in G_n \mid \langle A , G_{-1} \rangle = \{0\} \} = \{0\} \quad \text{for all } n \geq 0 , \tag{1.4}$$

c) *bitransitive* if in addition to (1.4) also

$$\{ A \in G_n \mid \langle A , G_1 \rangle = \{0\} \} = \{0\} \quad \text{for all } n \leq 0 \tag{1.5}$$

is fulfilled.

Note that in the special case n = 0 the equation in (1.4) (resp. in (1.5)) is valid if and only if the representation of G_0 in G_{-1} (resp. in G_1) is faithful.

Lemma 1

Let $G = \bigoplus_{n\geq -1} G_n$ be a transitive irreducible Z - graded Lie superalgebra. Suppose that $G_1 \neq \{0\}$ and that the representation of G_0 in G_1 is faithful. Then G is bitransitive.

Proof

As we have already mentioned our assumptions imply that the equation in (1.5) is valid for $n = 0$. Set

$$V = \{ A \in G_{-1} \mid \langle A , G_1 \rangle = \{0\} \} . \quad (1.6)$$

It is easy to see that V is a graded G_0 - submodule of G_{-1}. Since $G_1 \neq \{0\}$ the transitivity of G yields $\langle G_{-1} , G_1 \rangle \neq \{0\}$. It follows that $V \neq G_{-1}$ and hence that $V = \{0\}$.

Lemma 2

Let $G = \bigoplus_{n \geq -1} G_n$ be a transitive Z - graded Lie superalgebra. The Z - gradation of G is consistent with the Z_2 - gradation if and only if G_{-1} is an odd subspace of G (i.e. if and only if $G_{-1} \subset G_{\bar{1}}$).

Proof : By induction.

Lemma 3

Let $G = \bigoplus_{n \in Z} G_n$ be a Z - graded Lie superalgebra. Let E be a field containing K and let $E \underset{K}{\otimes} G = \bigoplus_{n \in Z} (E \underset{K}{\otimes} G_n)$ be the Z - graded Lie superalgebra over E which is obtained from G by an extension of the base field. Then $E \underset{K}{\otimes} G$ is transitive (resp. bitransitive) if and only if the same holds true for G.

The proof is straightforward and will be omitted.

Our interest in the transitivity relations (1.4) and (1.5) stems from the fact that in a sense (to be made precise in the corollary to proposition 1) they reduce the study of a Z - graded Lie superalgebra to the study of its "local part" $G_{-1} \oplus G_0 \oplus G_1$.

Proposition 1

Let $G = \bigoplus_{n \in Z} G_n$ and $G' = \bigoplus_{n \in Z} G'_n$ be two Z - graded Lie superalgebras.

Suppose we are given an even linear mapping

$$g : G_{-1} \oplus G_0 \oplus G_1 \longrightarrow G'_{-1} \oplus G'_0 \oplus G'_1 \tag{1.7}$$

which satisfies the following conditions :

a) $$g(G_{\pm 1}) = G'_{\pm 1} \quad , \quad g(G_0) \subset G'_0 \tag{1.8}$$

b) g is compatible with the multiplication in the sense that

$$g(\langle A , B \rangle) = \langle g(A) , g(B) \rangle \tag{1.9}$$

whenever $A \in G_n$, $B \in G_m$; $n , m , n+m \in \{-1, 0, 1\}$.

If the algebra G is generated by $G_{-1} \oplus G_0 \oplus G_1$ and if G' is bitransitive then there exists a unique extension of g to a homomorphism $\tilde{g} : G \longrightarrow G'$ of Z - graded Lie superalgebras.

Proof

We define inductively, for every integer $r \geq -1$, an even linear mapping

$$g_r : G_r \longrightarrow G'_r \tag{1.10}$$

such that for $r \geq 0$ the following conditions are fulfilled :

$$g_r(\langle A , B \rangle) = \langle g(A) , g_{r-1}(B) \rangle \quad \text{if } A \in G_1 , B \in G_{r-1} \tag{1.11}$$

$$g_r(\langle A , B \rangle) = \langle g(A) , g_r(B) \rangle \quad \text{if } A \in G_0 , B \in G_r \tag{1.12}$$

$$g_{r-1}(\langle A , B \rangle) = \langle g(A) , g_r(B) \rangle \quad \text{if } A \in G_{-1} , B \in G_r \; . \tag{1.13}$$

For $r \in \{-1, 0, 1\}$ we define g_r to be the linear mapping induced by g . By assumption the conditions (1.11) - (1.13) are then satisfied for $r = 0$ and $r = 1$.

Now let $r \geq 2$ and suppose that the mappings g_s with $-1 \leq s \leq r-1$ have already been defined. Let $p \geq 1$ be a positive integer and let

$$X_q \in G_1 \quad , \quad Y_q \in G_{r-1} \quad ; \quad 1 \leq q \leq p \; . \tag{1.14}$$

Using (1.9) and the induction hypothesis it is easy to see that for all $A \in G_{-1}$

$$\langle g(A) , \sum_{q=1}^{p} \langle g(X_q) , g_{r-1}(Y_q) \rangle \rangle = g_{r-1}(\langle A , \sum_{q=1}^{p} \langle X_q , Y_q \rangle \rangle) \; . \tag{1.15}$$

Since G' is bitransitive and since g_{-1} is surjective we conclude from (1.15) that the relation

$$\sum_{q=1}^{p}\langle X_q , Y_q\rangle = 0 \quad \text{implies} \quad \sum_{q=1}^{p}\langle g(X_q) , g_{r-1}(Y_q)\rangle = 0 . \qquad (1.16)$$

On the other hand, the algebra G is generated by $G_{-1} \oplus G_0 \oplus G_1$ and hence every element of G_r is of the form $\sum\langle X_q , Y_q\rangle$. Therefore, the equation

$$g_r\Big(\sum_{q=1}^{p}\langle X_q , Y_q\rangle\Big) = \sum_{q=1}^{p}\langle g(X_q) , g_{r-1}(Y_q)\rangle \qquad (1.17)$$

defines a mapping $g_r : G_r \longrightarrow G'_r$ which, obviously, is linear and even. We have to show that the conditions (1.11) - (1.13) are fulfilled. For (1.11) this is true by the very definition of g_r. To prove (1.12) and (1.13) we may assume that $B = \langle X , Y\rangle$ with $X \in G_1$, $Y \in G_{r-1}$; the equations then follow by a new application of (1.9) and of the induction hypothesis.

Obviously, our construction may also be carried through if the algebras G and G' are equipped with the inverted Z - gradations (see chapter 0, §2, 5)). Combining our results we obtain a family $(g_r)_{r\in Z}$ of even linear mappings

$$g_r : G_r \longrightarrow G'_r \qquad (1.18)$$

such that the equations (1.11) - (1.13) are fulfilled for all $r \in Z$. (Recall that for $r \in \{-1, 0, 1\}$ the mapping g_r is induced by g.)

It remains to show that

$$\langle g_r(A) , g_s(B)\rangle = g_{r+s}(\langle A , B\rangle) \qquad (1.19)$$

$$\text{for all } A \in G_r , B \in G_s \text{ and all } r , s \in Z .$$

The proof may be given by induction with respect to $|r|$. By construction of the family $(g_r)_{r\in Z}$ the equation (1.19) is valid if $r = 0$ or $|r| = 1$. Suppose now that $t \geq 0$ is some positive integer and that (1.19) is known for $|r| = t$. Let

$$X \in G_{\pm 1} , \quad Y \in G_{\pm t} , \quad B \in G_s . \qquad (1.20)$$

Using the induction hypothesis as well as the relations (1.11) - (1.13) it is not difficult to verify that

$$\langle g_{\pm(t+1)}(\langle X , Y\rangle) , g_s(B)\rangle = g_{\pm(t+1)+s}(\langle\langle X , Y\rangle, B\rangle) , \qquad (1.21)$$

and this implies (1.19) for $|r| = t+1$.

Now let

$$\tilde{g} : \bigoplus_{r\in Z} G_r \longrightarrow \bigoplus_{r\in Z} G'_r \qquad (1.22)$$

be the linear mapping defined by the family $(g_r)_{r\in Z}$. Then $\tilde{g}$ is a homomorphism of Z-graded Lie superalgebras which extends g. The uniqueness of such an extension is trivial.

Corollary

Using the notation introduced in proposition 1 let us assume that G and G' are bitransitive, that the algebra G is generated by $G_{-1} \oplus G_0 \oplus G_1$ and that the algebra G' is generated by $G'_{-1} \oplus G'_0 \oplus G'_1$. If g is bijective then $\tilde{g}$ is an isomorphism of Z-graded Lie superalgebras.

2. Various results on transitive Lie superalgebras

In this section $G = \bigoplus_{n\geqslant -1} G_n$ *denotes a finite-dimensional transitive irreducible consistently* Z*-graded Lie superalgebra.*

Lemma 4

The representation of the Lie algebra G_0 in G_{-1} is faithful and irreducible, hence G_0 is reductive.

Proposition 2 [26]

Suppose that $G_1 \neq \{0\}$. Then [20]

$$\langle G_0 , G_0\rangle \subset \langle G_{-1} , G_1\rangle . \qquad (1.23)$$

Proof

To begin with we remark that

$$\langle G_{-1}, \langle G_{-1}, G_1 \rangle\rangle = G_{-1} . \tag{1.24}$$

In fact, $\langle G_{-1}, \langle G_{-1}, G_1\rangle\rangle$ is a G_0 - invariant subspace of G_{-1}. This subspace is not equal to $\{0\}$, for otherwise the transitivity of G would imply that $\langle G_{-1}, G_1\rangle = \{0\}$ and hence that $G_1 = \{0\}$.

According to lemma 4 the Lie algebra G_0 is reductive, i.e. it is the direct product of its center with the semi-simple ideal $\langle G_0, G_0\rangle$. Obviously, $\langle G_{-1}, G_1\rangle$ is an ideal of G_0; let S be the centralizer of $\langle G_{-1}, G_1\rangle$ in G_0. Since G_0 is reductive the relation (1.23) will follow if we can prove that S is abelian.

Let $A, B \in S$; we have to show that $\langle A, B\rangle = 0$. In view of the transitivity of G and of equation (1.24) this will be the case if

$$\langle\langle A, B\rangle, \langle G_{-1}, \langle G_{-1}, G_1\rangle\rangle\rangle = \{0\} . \tag{1.25}$$

But if $X, Y \in G_{-1}$ and $U \in G_1$ we have

$$\begin{aligned} \langle A, \langle B, \langle X, \langle Y, U\rangle\rangle\rangle\rangle &= \langle A, \langle\langle B, X\rangle, \langle Y, U\rangle\rangle\rangle \\ &= -\langle A, \langle Y, \langle\langle B, X\rangle, U\rangle\rangle\rangle \\ &= -\langle\langle A, Y\rangle, \langle\langle B, X\rangle, U\rangle\rangle \end{aligned} \tag{1.26}$$

(recall that $\langle G_{-1}, G_{-1}\rangle = \{0\}$).

Evidently, the left hand side is skew-symmetric in X, Y whereas the right hand side changes sign under the simultaneous interchange $A \leftrightarrow B$ and $X \leftrightarrow Y$. It follows that our expression is symmetric in A, B, which implies (1.25).

Proposition 3

We suppose that the field K is algebraically closed.

a) Let G_0^0 be the center of the (reductive) Lie algebra G_0. Then

$$\dim G_0^0 \leq 1 \tag{1.27}$$

and in the case $\dim G_0^0 = 1$ there exists a unique element C in G_0^0 such that for all integers $n \geq -1$

$$\langle C, X\rangle = nX \quad \text{if } X \in G_n . \tag{1.28}$$

b) For every integer $n \geq -1$ the representation of G_0 in G_n is completely reducible.

c) If G_0 is abelian we have $G_n = \{0\}$ for all $n \geq 1$.

Proof

a) Consider the representation of G_0 in G_{-1}; it is faithful and irreducible, hence the elements of G_0^o are represented by scalar multiples of the identity. This implies (1.27); furthermore, if $G_0^o \neq \{0\}$, there exists a unique element $C \in G_0^o$ such that

$$\langle C, X\rangle = -X \quad \text{for all } X \in G_{-1} . \tag{1.29}$$

This element C has the desired property. In fact, suppose that (1.28) is valid for some integer $n \geq -1$. If $X \in G_{n+1}$ it is easy to check that

$$\langle Y, \langle C, X\rangle - (n+1)X\rangle = 0 \quad \text{for all } Y \in G_{-1} . \tag{1.30}$$

Since G is transitive this implies

$$\langle C, X\rangle = (n+1)X , \tag{1.31}$$

as required.

b) follows from a) and the fact that G_0 is reductive.

c) Suppose that G_0 is abelian. Since the representation of G_0 in G_{-1} is irreducible we conclude that $\dim G_{-1} = 1$. Let X be some non-zero element of G_{-1}. Then we have for all integers $n \geq 1$

$$2\langle X, \langle X, G_n\rangle\rangle = \langle\langle X, X\rangle, G_n\rangle = \{0\} . \tag{1.32}$$

Using the transitivity of G we conclude that $\langle X, G_n\rangle = \{0\}$, hence that $G_n = \{0\}$.

Proposition 4

We suppose that the field K is algebraically closed and that the representation of G_0 in G_1 is irreducible (in particular, $G_1 \neq \{0\}$).

Let h be a Cartan subalgebra of the (reductive) Lie algebra G_0. We choose a fundamental system of simple roots of G_0 with respect to h.

Let λ (resp. μ) be the highest (resp. lowest) weight of the representation of G_0 in G_{-1} (resp. in G_1) and let $X_\lambda \in G_{-1}$ (resp. $Y_\mu \in G_1$) be a weight vector associated with it.

We distinguish two cases.

a) If the representations of G_0 in G_{-1} and G_1 are contragredient to each other then :

1) $$\mu = -\lambda \tag{1.33}$$

2) $$\langle X_\lambda , Y_\mu \rangle = H \tag{1.34}$$

where H is some non-zero element of h which does not belong to the center of G_0 .

b) If the representations of G_0 in G_{-1} and G_1 are not contragredient to each other then :

1) $$\langle X_\lambda , Y_\mu \rangle = E_\alpha \tag{1.35}$$

where $\alpha = \lambda + \mu$ is a root of the Lie algebra G_0 and E_α is a root vector associated with it.

2) $$\langle G_{-1} , G_1 \rangle = \langle G_0 , G_0 \rangle \tag{1.36}$$

3) The Lie algebra $\langle G_0 , G_0 \rangle$ is simple.

Proof

Let $\alpha_1, \ldots, \alpha_m$ be the fundamental system of simple roots of G_0 with respect to h which has been chosen. For every root γ of G_0 let F_γ be a root vector associated with it. (Recall that $F_\gamma \in \langle G_0 , G_0 \rangle$.)

The vector space G_{-1} is generated by the vectors of the form

$$\langle F_{-\gamma_1} , \langle F_{-\gamma_2} , \ldots \langle F_{-\gamma_r} , X_\lambda \rangle \ldots \rangle\rangle \tag{1.37}$$

with $\gamma_1, \ldots, \gamma_r \in \{\alpha_1, \ldots, \alpha_m\}$

and the vector space G_1 is generated by the vectors of the form

$$\langle F_{\delta_1} , \langle F_{\delta_2} , \ldots \langle F_{\delta_s} , Y_\mu \rangle \ldots \rangle\rangle \tag{1.38}$$

with $\delta_1, \ldots, \delta_s \in \{\alpha_1, \ldots, \alpha_m\}$.

It follows that the vector space $\langle G_{-1} , G_1 \rangle$ is generated by the vectors of the form

$$\langle F_{\beta_1}, \langle F_{\beta_2}, \dots \langle F_{\beta_t}, \langle X_\lambda, Y_\mu \rangle\rangle \dots \rangle\rangle \tag{1.39}$$

with $\beta_1, \dots, \beta_t \in \{\alpha_1, \dots, \alpha_m, -\alpha_1, \dots, -\alpha_m\}$.

Consequently, $\langle X_\lambda, Y_\mu \rangle$ is a cyclic vector for the representation of $\langle G_0, G_0 \rangle$ in $\langle G_{-1}, G_1 \rangle$.

From proposition 2 and proposition 3,c) we know that $\langle G_0, G_0 \rangle$ is a non-zero semi-simple Lie algebra which is contained in $\langle G_{-1}, G_1 \rangle$. Hence $\langle X_\lambda, Y_\mu \rangle$ cannot belong to the center of G_0 , which implies, in particular, that

$$\langle X_\lambda, Y_\mu \rangle \neq 0 \; . \tag{1.40}$$

Evidently, the vector $\langle X_\lambda, Y_\mu \rangle$ belongs to the weight $\lambda + \mu$ of the adjoint representation of G_0 .

The representations of G_0 in G_{-1} and G_1 are contragredient to each other if and only if $\lambda + \mu = 0$. If this is the case we conclude that $\langle X_\lambda, Y_\mu \rangle$ lies in h and part a) of our proposition is proved.

Suppose now that the representations of G_0 in G_{-1} and G_1 are not contragredient to each other. Set

$$\alpha = \lambda + \mu \; ; \tag{1.41}$$

then α is a root of G_0 and $\langle X_\lambda, Y_\mu \rangle$ is a root vector associated with it. Let $\tilde{G}_0$ be the simple ideal of the Lie algebra G_0 to which the root α belongs. Then we have $\langle X_\lambda, Y_\mu \rangle \in \tilde{G}_0$ and the above results, combined with proposition 2 , imply that

$$\langle G_{-1}, G_1 \rangle \subset \tilde{G}_0 \subset \langle G_0, G_0 \rangle \subset \langle G_{-1}, G_1 \rangle \; . \tag{1.42}$$

This proves the statements 2) and 3) of part b).

Proposition 5

We suppose that the field K is algebraically closed and that $G_1 \neq \{0\}$ but $G_n = \{0\}$ if $n \geqslant 2$.

Then the representation of G_0 in G_1 is irreducible; furthermore, this representation is faithful provided that $\dim G_1 \geqslant 2$.

Proof

To begin with we prove the following statement :

If

$$\langle\langle G_0 , G_0 \rangle , G_1 \rangle = \{0\} , \tag{1.43}$$

it follows that $\dim G_1 = 1$.

In fact, let A be a non-zero element of G_1 . The equation (1.43) and proposition 3,a) imply that $G_{-1} \oplus G_0 \oplus K\cdot A$ is a subalgebra of G . The representations of G_0 in G_{-1} and $K\cdot A$ are not contragredient to each other (see proposition 3), hence we derive from proposition 4 that

$$\langle G_0 , G_0 \rangle = \langle A , G_{-1} \rangle \tag{1.44}$$

is a simple Lie algebra.

Consider the linear mapping

$$f_A : G_{-1} \longrightarrow G_0 \tag{1.45,a}$$

which is defined by

$$f_A(Y) = \langle A , Y \rangle \quad \text{if } Y \in G_{-1} . \tag{1.45,b}$$

This mapping is $\langle G_0 , G_0 \rangle$ - invariant and maps G_{-1} onto $\langle G_0 , G_0 \rangle$.

Since G_{-1} and $\langle G_0 , G_0 \rangle$ are simple $\langle G_0 , G_0 \rangle$ - modules any two $\langle G_0 , G_0 \rangle$ - invariant linear mappings of G_{-1} into $\langle G_0 , G_0 \rangle$ are proportional.

Now let A_1 , A_2 be two non-zero elements of G_1 . Using the above results as well as the transitivity of G it follows immediately that A_1 and A_2 are proportional. Thus our statement is proved.

We next show that the representation of G_0 in G_1 is irreducible. According to proposition 3 this representation is completely reducible. Suppose that there exists a decomposition

$$G_1 = G_1' \oplus G_1'' \tag{1.46}$$

of the G_0 - module G_1 into the direct sum of two non-zero G_0 - submodules G_1' and G_1'' . We apply proposition 2 to the subalgebra $G_{-1} \oplus G_0 \oplus G_1'$ of G and obtain

$$\langle G_0 , G_0 \rangle \subset \langle G_{-1} , G_1' \rangle . \tag{1.47}$$

It follows that

$$\langle\langle G_0, G_0\rangle, G_1''\rangle \subset \langle\langle G_{-1}, G_1'\rangle, G_1''\rangle \subset \langle G_1', \langle G_{-1}, G_1''\rangle\rangle \subset G_1' . \quad (1.48)$$

Since G_1'' is G_0 - invariant we conclude that

$$\langle\langle G_0, G_0\rangle, G_1''\rangle = \{0\} . \quad (1.49)$$

An analogous argument leads to

$$\langle\langle G_0, G_0\rangle, G_1'\rangle = \{0\} , \quad (1.50)$$

hence we have shown that equation (1.43) holds true. Now the first part of our proof yields $\dim G_1 = 1$, whereas the decomposability of G_1 implies $\dim G_1 \geq 2$. Thus we have arrived at a contradiction.

Finally, let us assume that the representation of G_0 in G_1 is not faithful. Recalling proposition 3,a) we conclude from this that there exists a simple ideal J of $\langle G_0, G_0\rangle$ such that

$$\langle J, G_1\rangle = \{0\} . \quad (1.51)$$

Let $A \in G_1$; we consider again the mapping $f_A : G_{-1} \longrightarrow G_0$ which has been defined in (1.45) . It follows from (1.51) that f_A is J - invariant. Since the G_0 - module G_{-1} is faithful and irreducible, we know that the J - module G_{-1} is the direct sum of faithful irreducible J - submodules (which are isomorphic to each other). Evidently, f_A maps each of these submodules into J . It follows that

$$f_A(G_{-1}) \subset J \quad \text{for all } A \in G_1 \quad (1.52)$$

which is to say that

$$\langle G_1, G_{-1}\rangle \subset J . \quad (1.53)$$

In view of proposition 2 this implies

$$J = \langle G_0, G_0\rangle . \quad (1.54)$$

But then the first part of our proof yields $\dim G_1 = 1$.

3. Construction of two types of transitive Lie superalgebras

The results of the foregoing section suggest the construction of two types of transitive Z - graded Lie superalgebras.

a) The first construction has its origin in proposition 3. Let $G = \bigoplus_{n\in Z} G_n$ be an arbitrary consistently Z-graded Lie superalgebra and let G'_0 be a one-dimensional vector space. We choose some non-zero element C in G'_0 and define a new consistently Z-graded Lie superalgebra $G^Z = \bigoplus_{n\in Z} G^Z_n$ as follows. Set

$$G^Z_0 = G_0 \oplus G'_0 \tag{1.55,a}$$

$$G^Z_n = G_n \quad \text{if } n \in Z,\ n \neq 0 \ . \tag{1.55,b}$$

On the Z-graded vector space G^Z there exists a unique algebra structure such that G is a subalgebra of G^Z and such that

$$\langle C , X\rangle = -\langle X , C\rangle = nX \quad \text{for all } X \in G^Z_n \text{ and all } n \in Z \ . \tag{1.56}$$

Equipped with this structure G^Z is a consistently Z-graded Lie superalgebra (see chapter III, §1, n°2).

Suppose now that $G_{-1} \neq \{0\}$ and that the center of the Lie algebra G_0 is trivial. Then it is easy to see that G^Z is transitive if and only if G is transitive. Furthermore, the propositions 2 and 3 imply the following lemma.

<u>Lemma</u> 5

We suppose that the field K is algebraically closed.

Let $G = \bigoplus_{n\geq -1} G_n$ be a finite-dimensional transitive irreducible consistently Z-graded Lie superalgebra such that $G_1 \neq \{0\}$. We define a Z-graded subalgebra $\tilde{G}$ of G by

$$\tilde{G} = G_{-1} \oplus \langle G_{-1}, G_1\rangle \oplus \bigoplus_{n\geq 1} G_n \ . \tag{1.57}$$

This algebra is again transitive and irreducible. Moreover, there are only the following two possibilities.

1) $\langle G_{-1}, G_1\rangle = G_0$ and hence $\tilde{G} = G$.

2) $\langle G_{-1}, G_1\rangle = \langle G_0, G_0\rangle \neq G_0$ and the Z-graded Lie superalgebras G and $\tilde{G}^Z$ are isomorphic. (Recall that $\langle G_0, G_0\rangle$ is semi-simple and non-zero.)

b) The other construction to be discussed in this section is derived from proposition 5. Let H be an arbitrary Lie algebra. We set

$$G_{-1} = H \ , \quad G_0 = H \ , \quad G_1 = K \qquad (1.58,a)$$

and consider the Z - graded vector space

$$G = G_{-1} \oplus G_0 \oplus G_1 \ . \qquad (1.58,b)$$

Then there exists on G a unique structure of a consistently Z - graded Lie superalgebra such that

$$\langle Q , a \rangle = 0 \qquad (1.59,a)$$

$$\langle Q , Q' \rangle = [Q , Q'] \qquad (1.59,b)$$

$$\langle Q , U \rangle = [Q , U] \qquad (1.59,c)$$

$$\langle a , U \rangle = a U \qquad (1.59,d)$$

for all $Q , Q' \in G_0$; $U \in G_{-1}$; $a \in G_1$.

The algebra which has been defined will be denoted by H^{ξ}. (Our notation in (1.58) and (1.59) is somewhat oversimplified, however, the meaning of these formulae should be obvious.)

Suppose that $H \neq \{0\}$. Then the algebra H^{ξ} is transitive if and only if the center of H is trivial.

Lemma 6

We suppose that the field K is algebraically closed.

Let $G = \bigoplus_{n \geq -1} G_n$ be a finite-dimensional transitive irreducible consistently Z - graded Lie superalgebra such that $\dim G_1 = 1$.

Then $\langle G_0 , G_0 \rangle$ is a simple Lie algebra and the Z - graded Lie superalgebra G is isomorphic either to $\langle G_0 , G_0 \rangle^{\xi}$ or to $(\langle G_0 , G_0 \rangle^{\xi})^{Z}$.

Conversely, let H be any (finite-dimensional) simple Lie algebra. Then H^{ξ} and $(H^{\xi})^{Z}$ have the properties which we have assumed to hold for G.

Proof

The second half of the lemma is obvious.

To verify the first half it is sufficient to prove the following result (see lemma 5).

Let us assume in addition that

$$\langle G_{-1}, G_1 \rangle = G_0 \ . \tag{1.60}$$

Then the Lie algebra G_0 is simple and G is isomorphic to $(G_0)^{\xi}$.

The representations of G_0 in G_{-1} and G_1 are not contragredient to each other (see proposition 3), hence we conclude from proposition 4 that G_0 is a simple Lie algebra. It follows that the representation of G_0 in G_1 is trivial.

Let A be any non-zero element of G_1 . Consider the linear mapping

$$f : G_{-1} \longrightarrow G_0 \tag{1.61,a}$$

which is defined by

$$f(Y) = \langle A, Y \rangle \quad \text{for all } Y \in G_{-1} \ . \tag{1.61,b}$$

Evidently, f is G_0 - invariant, and the transitivity of G implies that $f \neq 0$. Since both G_{-1} and G_0 are simple G_0 - modules we conclude that f is an isomorphism of G_0 - modules.

Let us next prove that $G_n = \{0\}$ if $n \geq 2$. Because of the transitivity of G it is sufficient to show that $G_2 = \{0\}$. Let T be any element of G_2 . Then the equation

$$\langle Y, T \rangle = \lambda(Y)\, A \quad ; \quad Y \in G_{-1} \tag{1.62}$$

defines a linear form λ on G_{-1} and we have

$$2\,\lambda(Y)\, f(Y) = 2\langle Y, \langle Y, T\rangle\rangle = \langle\langle Y, Y\rangle, T\rangle = 0 \tag{1.63}$$

for all $Y \in G_{-1}$. We conclude that $\lambda = 0$ and hence (using the transitivity once again) that $T = 0$.

It is now easy to see that the Z - graded Lie superalgebras G and $(G_0)^{\xi}$ are isomorphic.

4. Filtration of Lie superalgebras

The following definition of a filtration is somewhat more restrictive than usual.

Definition 3

A Lie superalgebra L is called *filtered* if we are given a family $(L^n)_{n\in Z}$ of Z_2 - graded subspaces of L such that

$$L^n \supset L^m \quad \text{if } n \leqslant m \tag{1.64}$$

$$L^n = L \quad \text{if } n \leqslant -1 \tag{1.65}$$

$$\bigcap_{n\in Z} L^n = \{0\} \tag{1.66}$$

$$\langle L^n , L^m \rangle \subset L^{n+m} \quad \text{for all } n , m \in Z \ . \tag{1.67}$$

The filtration $(L^n)_{n\in Z}$ (and the algebra L itself) will be called *transitive* if

$$L^{n+1} = \{ A \in L^n \mid \langle A , L \rangle \subset L^n \} \quad \text{for all } n \geqslant 0 \ . \tag{1.68}$$

From (1.67) we deduce that $\langle L^0 , L^n \rangle \subset L^n$ for all $n \in Z$; hence L^0 is a graded subalgebra of L and the adjoint representation of L induces a natural graded representation of L^0 in L^n , for all $n \in Z$.

It follows from the equations (1.65) and (1.68) that a transitive filtration $(L^n)_{n\in Z}$ is fixed once we are given the graded subalgebra L^0 of L . Conversely :

Lemma 7

Let L be a Lie superalgebra and let L^0 be a graded subalgebra of L . Define the family $(L^n)_{n\in Z}$ of subspaces of L by the equations (1.65) and (1.68). Then the L^n are Z_2 - graded subspaces of L and the relations (1.64) and (1.67) are fulfilled. Moreover, the condition (1.66) is satisfied if and only if L^0 does not contain any non-zero graded ideal of L .

Proof

The relation (1.64) is obvious, furthermore, it is easy to see that the L^n are Z_2 - graded subspaces of L .

Comment on (1.67). By definition, this relation is fulfilled if $n \leqslant -1$ or $m \leqslant -1$. It remains to prove (1.67) for the cases where $n , m \geqslant 0$, and this can be done by induction with respect to $n+m$.

Comment on (1.66). It follows from (1.67) that $\bigcap_{n\in Z} L^n$ is a graded ideal of L which, obviously, is contained in L^0. On the other hand, if J is any graded ideal of L which is contained in L^0 then $J \subset L^n$ for all $n \in Z$.

Let L be a filtered Lie superalgebra and let $(L^n)_{n\in Z}$ be its filtration. Then we can define the Z - graded algebra

$$\operatorname{gr} L = \bigoplus_{n\in Z} \operatorname{gr}_n L \tag{1.69}$$

associated with L. Recall that

$$\operatorname{gr}_n L = L^n / L^{n+1} \tag{1.70}$$

and that the product mapping $(\operatorname{gr}_n L) \times (\operatorname{gr}_m L) \longrightarrow \operatorname{gr}_{n+m} L$ is obtained from the product mapping $L^n \times L^m \longrightarrow L^{n+m}$ by going to the quotients. The spaces $\operatorname{gr}_n L$ are Z_2 - graded. It is now easy to see that gr L is a Z - graded Lie superalgebra. Of course we have

$$\operatorname{gr}_n L = \{0\} \quad \text{if } n \leqslant -2 . \tag{1.71}$$

The Z - gradation of gr L is consistent with the Z_2 - gradation if and only if

$$L_{\bar 0}^{2r-1} = L_{\bar 0}^{2r} , \quad L_{\bar 1}^{2r} = L_{\bar 1}^{2r+1} \quad \text{for all } r \in Z . \tag{1.72}$$

Lemma 8

Let L be a filtered Lie superalgebra (with the filtration $(L^n)_{n\in Z}$).

a) The filtered Lie superalgebra L is transitive if and only if the Z - graded Lie superalgebra gr L is transitive.

b) Suppose that L is transitive. The Z - gradation of gr L is consistent with the Z_2 - gradation if and only if $L_{\bar 0} \subset L^0$.

c) Suppose that $L_{\bar 0} \subset L^0$. The Z - graded Lie superalgebra gr L is irreducible if and only if $L_{\bar 1}^0$ is a maximal (proper) $L_{\bar 0}$ - invariant subspace of $L_{\bar 1}$.

The proof of the lemma is straightforward and will be omitted (concerning part b) see lemma 2).

Now let $L = \bigoplus_{n \geq -1} G_n$ be a Z-graded Lie superalgebra. Let us define

$$L^n = \bigoplus_{m \geq n} G_m \quad \text{for all } n \in Z . \tag{1.73}$$

It is easy to see that $(L^n)_{n \in Z}$ is a filtration of L; this filtration is said to be associated with the given Z-gradation of L. Obviously, the Z-graded Lie superalgebra gr L which is associated with the filtration $(L^n)_{n \in Z}$ is canonically isomorphic with the original Z-graded Lie superalgebra L.

In the subsequent paragraphs of this chapter it will be important to know whether, conversely, a filtration stems from a Z-gradation in the way described above. The following proposition contains a simple criterion for this to be the case.

Proposition 6

We suppose that the field K is algebraically closed.

Let L be a finite-dimensional transitive filtered Lie superalgebra (equipped with the filtration $(L^n)_{n \in Z}$). Suppose that gr L is an irreducible consistently Z-graded Lie superalgebra such that the center of $gr_0 L$ is non-trivial.

Then there exists a Z-gradation $L = \bigoplus_{n \geq -1} G_n$ of the Lie superalgebra L which induces the given filtration $(L^n)_{n \in Z}$. Consequently, the Lie superalgebras L and gr L are isomorphic.

(Note that the Z-gradation $(G_n)_{n \in Z}$ is consistent with the Z_2-gradation of L.)

Proof

According to our assumptions gr L is a transitive irreducible consistently Z-graded Lie superalgebra and the center of $gr_0 L$ is non-trivial. Hence (see proposition 3) there exists a unique element $\tilde{C}$ of the center of $gr_0 L$ such that for all integers n

$$\langle \tilde{C}, \tilde{X} \rangle = n\tilde{X} \quad \text{if } \tilde{X} \in gr_n L \ . \tag{1.74}$$

Let C be an element of $L^0_{\bar{0}}$ whose image under the canonical mapping $L^0 \longrightarrow L^0/L^1 = gr_0 L$ is equal to $\tilde{C}$. It follows from (1.74) that every eigenvalue of ad C is an integer $n \geq -1$.

Let us consider the primary decomposition of L with respect to ad C. If we define for every integer n

$$G_n = \{ A \in L \mid (\mathrm{ad}\,C - n)^r(A) = 0 \text{ if } r \text{ is sufficiently large} \} \tag{1.75}$$

it follows that

$$L = \bigoplus_{n \geq -1} G_n \ . \tag{1.76}$$

Obviously, the G_n are Z_2 - graded subspaces of L. On the other hand the product mapping $L \times L \longrightarrow L$ is $L_{\bar{0}}$ - invariant; in particular, it is invariant under ad C. As is well-known [12] this implies

$$\langle G_n , G_m \rangle \subset G_{n+m} \quad \text{for all } n, m \in Z \ . \tag{1.77}$$

Thus $(G_n)_{n \in Z}$ is a Z - gradation of the Lie superalgebra L.

It remains to show that

$$L^n = \bigoplus_{m \geq n} G_m \quad \text{for all } n \geq -1 \ . \tag{1.78}$$

Let n be any integer and let

$$g_n : L^n \longrightarrow L^n / L^{n+1} = gr_n L \tag{1.79}$$

be the canonical mapping. Evidently,

$$g_n(\langle C, X \rangle) = \langle \tilde{C}, g_n(X) \rangle = n\, g_n(X) \quad \text{for all } X \in L^n \ . \tag{1.80}$$

It follows [12] that

$$g_n(G_m \cap L^n) = \{0\} \quad \text{if } m \in Z,\ m \neq n \ , \tag{1.81}$$

i.e. that

$$G_m \cap L^n \subset L^{n+1} \quad \text{if } m, n \in Z,\ m \neq n \ . \tag{1.82}$$

Bearing in mind the equation (1.76) it is now not difficult to show that (1.78) is valid.

§ 2 Some general properties of simple Lie superalgebras

We remind the reader that from this paragraph on we shall assume that all Lie superalgebras are *finite-dimensional*.

1. Some elementary results on simple Lie superalgebras [6]

Definition 1

A Lie superalgebra L is called *simple* if it does not have any graded ideals which are different from {0} and L and if, moreover, $\langle L , L \rangle \neq \{0\}$.

Remarks

1) Recall that a left or right graded ideal of L is automatically a two-sided ideal.

2) The condition $\langle L , L \rangle \neq \{0\}$ serves to eliminate the zero-dimensional and the two one-dimensional Lie superalgebras. It follows that $\langle L , L \rangle$ is equal to L (see chapter I, §1, example 1)).

According to definition 1 it is conceivable that a simple Lie superalgebra might contain non-trivial *non-graded* ideals. Actually, however, this is not the case.

Proposition 1

A simple Lie superalgebra L does not have any left or right ideals (graded or not) except for {0} and L .

Proof

The linear mapping

$$\gamma : L \longrightarrow L \qquad (2.1,a)$$

defined by

$$\gamma(A) = (-1)^{\alpha} A \quad \text{if } A \in L_{\alpha} , \alpha \in Z_2 \qquad (2.1,b)$$

is an automorphism of the Lie superalgebra L . If B is any element of L its homogeneous component of degree $\beta \in Z_2$ is equal to $\frac{1}{2}(B + (-1)^{\beta}\gamma(B))$.

In particular, a subspace of L is Z_2 - graded if and only if it is invariant under γ .

Let J be a left ideal of L which is different from {0} and L . Then $\gamma(J)$ is also a left ideal of L . Consequently, $J + \gamma(J)$ and $J \cap \gamma(J)$ are graded ideals of L and hence

$$J + \gamma(J) = L \quad , \quad J \cap \gamma(J) = \{0\} \ . \tag{2.2}$$

Thus the vector space L is the direct sum of its subspaces J and $\gamma(J)$; moreover, it follows that

$$L_\alpha = \{ B + (-1)^\alpha \gamma(B) \mid B \in J \} \quad \text{if } \alpha \in Z_2 \ . \tag{2.3}$$

Let

$$\tau : L \longrightarrow L \tag{2.4,a}$$

be the linear mapping which is defined by

$$\tau(B) = B \quad , \quad \tau(\gamma(B)) = -\gamma(B) \quad \text{for all } B \in J \ . \tag{2.4,b}$$

Then we have

$$\tau^2 = \text{id} \tag{2.5}$$

and

$$\tau(L_{\bar{0}}) = L_{\bar{1}} \quad , \quad \tau(L_{\bar{1}}) = L_{\bar{0}} \ . \tag{2.6}$$

Furthermore, the fact that J and $\gamma(J)$ are left ideals implies that τ commutes with the adjoint representation :

$$\tau(\langle A , B \rangle) = \langle A , \tau(B) \rangle \quad \text{for all } A , B \in L \ . \tag{2.7}$$

But according to the subsequent lemma, a mapping τ with these properties does not exist.

The case of a right ideal J is treated similarly.

Lemma 1

Let L be a simple Lie superalgebra. If τ is an odd linear mapping of L into itself such that

$$\tau(\langle A , B \rangle) = \langle A , \tau(B) \rangle \quad \text{for all } A , B \in L \ , \tag{2.8}$$

then $\tau = 0$.

Proof

It is easy to see that the kernel and the image of τ are graded ideals of L, hence either $\tau = 0$ or else τ is bijective.

Suppose that τ is bijective. Let A and B be any two homogeneous elements of L. If A and B have the same degree then

$$\langle \tau(A), \tau(B)\rangle = -\tau^2(\langle B, A\rangle) . \tag{2.9}$$

But one side of this equation is symmetric in A, B and the other is skew-symmetric. Thus $\langle A, B\rangle = 0$ if A and B are homogeneous of the same degree.

On the other hand, if A and B are homogeneous of different degrees then A and $\tau(B)$ are homogeneous of the same degree, hence equation (2.8) and our previous result imply that $\langle A, B\rangle = 0$ in this case, too. Thus we have shown that $\langle L, L\rangle = \{0\}$, a contradiction.

Lemma 2

Let L be a simple Lie superalgebra. Then [20]

1)
$$\langle L_{\bar{0}}, L_{\bar{1}}\rangle = L_{\bar{1}} \tag{2.10}$$

2) If $L_{\bar{1}} \neq \{0\}$ we have

$$\langle L_{\bar{1}}, L_{\bar{1}}\rangle = L_{\bar{0}} \tag{2.11}$$

$$\{A \in L \mid \langle A, L_{\bar{1}}\rangle = \{0\}\} = \{0\} . \tag{2.12}$$

In particular, the adjoint representation ad' of $L_{\bar{0}}$ in $L_{\bar{1}}$ is faithful (see chapter I, §1, definition 3).

3) If $A \longrightarrow A_V$ is a graded representation of L in some finite-dimensional graded vector space V then

$$\mathrm{str}(A_V) = 0 \quad \text{for all } A \in L \tag{2.13}$$

(see chapter I, equation (3.50)).

Proof

1) and 2) follow from the fact that $L_{\bar{0}} \oplus \langle L_{\bar{0}}, L_{\bar{1}}\rangle$, $\langle L_{\bar{1}}, L_{\bar{1}}\rangle \oplus L_{\bar{1}}$ and $\{A \in L \mid \langle A, L_{\bar{1}}\rangle = \{0\}\}$ are graded ideals of L. Statement 3) is valid for every Lie superalgebra L such that $\langle L, L\rangle = L$ (see chapter I, equation (3.51)).

Proposition 2

Let L be a simple Lie superalgebra.

1) An invariant bilinear form (see chapter I, §3, n°3, example 4,c)) on L is either non-degenerate or equal to zero.

2) Every invariant bilinear form on L is supersymmetric (see chapter I, §3, n°3, definition 4).

3) The invariant bilinear forms on L are either all even or else all odd.

4) If the field K is algebraically closed then all invariant bilinear forms on L are proportional to each other.

Proof

1) Let ψ be any invariant bilinear form on L . The subspace

$$J = \{ B \in L \mid \psi(A,B) = 0 \text{ for all } A \in L \} \tag{2.14}$$

is a left (a priori not necessarily graded) ideal of L . Hence proposition 1 proves our statement.

2) This result is valid for all Lie superalgebras L such that $\langle L , L \rangle = L$ (see §3, proposition 1).

3) The homogeneous components of an invariant bilinear form are again invariant. Therefore, bearing in mind part 1) of this proposition, it is sufficient to prove :

Let ψ and ψ' be two invariant bilinear forms on L which are homogeneous of different degrees. If ψ is non-degenerate then $\psi' = 0$.

In fact, since ψ is non-degenerate there exists a unique linear mapping τ of L into itself such that

$$\psi'(A,B) = \psi(A,\tau(B)) \text{ for all } A , B \in L \, . \tag{2.15}$$

It is easy to see that τ satisfies the conditions of lemma 1 . Hence we conclude that $\tau = 0$.

4) is an easy consequence of 1).

Remark 3)

Contrary to what is known for simple Lie algebras it is not at all true that on every simple Lie superalgebra there exists a non-degenerate invariant bilinear form. This defect is the origin of the main difficulties in proving the classification theorem for simple Lie superalgebras.

To conclude this section we give some elementary criteria for a Lie superalgebra to be simple.

Lemma 3

Let L be a Lie superalgebra which satisfies the following conditions :

1) The representation of $L_{\bar{0}}$ in $L_{\bar{1}}$ is faithful and irreducible.

2)
$$\langle L_{\bar{1}}, L_{\bar{1}} \rangle = L_{\bar{0}} \ . \tag{2.16}$$

Then the Lie superalgebra L is simple, provided that $L_{\bar{0}} \neq \{0\}$.

The proof is obvious.

Lemma 4

Let $G = \bigoplus_{n \geq -1} G_n$ be a (not necessarily consistently) Z - graded Lie superalgebra such that $G_{-1} \neq \{0\}$.

a) If the Lie superalgebra G is simple, then :

1) G is transitive and irreducible,

2)
$$\langle G_{-1}, G_1 \rangle = G_0 \tag{2.17}$$

and in particular

3)
$$G_0 \neq \{0\} \quad , \quad G_1 \neq \{0\} \ . \tag{2.18}$$

b) Conversely, suppose that $G_1 \neq \{0\}$ and that G satisfies the condition 1) as well as the following one :

4)
$$\langle G_n, G_1 \rangle = G_{n+1} \quad \text{for all } n \geq -1 \ . \tag{2.19}$$

Then the Lie superalgebra G is simple.

Proof

a) Let V be a Z_2 - graded subspace of G such that

$$\langle G_{-1}, V \rangle \subset V \quad , \quad \langle G_0, V \rangle \subset V \ . \tag{2.20}$$

We set

$$G^{+} = \bigoplus_{n\geq 1} G_n \qquad (2.21)$$

and define for all positive integers $n \geq 0$

$$V^n = \underbrace{\langle G^{+}, \langle G^{+}, \ldots . \langle G^{+}}_{n \text{ factors } G^{+}}, V\rangle ..\rangle\rangle \quad . \qquad (2.22)$$

Then it is easy to see that

$$\tilde{V} = \sum_{n\geq 0} V^n \qquad (2.23)$$

is a graded ideal of the Lie superalgebra G .

Suppose now that G is simple. If we choose

$$V = \{ A \in \bigoplus_{n\geq 0} G_n \mid \langle A , G_{-1}\rangle = \{0\} \} \qquad (2.24)$$

the ideal $\tilde{V}$ is contained in $\bigoplus_{n\geq 0} G_n$. It follows that $\tilde{V} = \{0\}$ and hence that G is transitive.

On the other hand, let V be a non-zero graded G_0 - submodule of G_{-1} . Then $\tilde{V}$ is non-zero and is contained in $V \oplus \bigoplus_{n\geq 0} G_n$. This implies that $V = G_{-1}$; thus G is irreducible.

The equation (2.17) follows from the fact that $G_{-1} \oplus \langle G_{-1} , G_1\rangle \oplus G^{+}$ is a graded ideal of G .

Part b) is obvious.

2. Discussion of the $L_{\bar{0}}$ - module $L_{\bar{1}}$ [6]

In this section we shall investigate the adjoint representation ad' of $L_{\bar{0}}$ in $L_{\bar{1}}$. This discussion will draw our attention to an interesting special class of simple Lie superalgebras, the so-called classical ones.

Proposition 3

Let L be a simple Lie superalgebra. Suppose that $L_{\bar{1}}$ is the sum

$$L_{\bar{1}} = L_{\bar{1}}^1 + L_{\bar{1}}^2 \qquad (2.25)$$

of two proper $L_{\bar{0}}$ - invariant subspaces $L_{\bar{1}}^{1}$ and $L_{\bar{1}}^{2}$.

Then this sum is direct, i.e.

$$L_{\bar{1}}^{1} \cap L_{\bar{1}}^{2} = \{0\} \tag{2.26}$$

and the $L_{\bar{0}}$ - modules $L_{\bar{1}}^{1}$ and $L_{\bar{1}}^{2}$ are irreducible. Furthermore, we have

$$\langle L_{\bar{1}}^{1}, L_{\bar{1}}^{1} \rangle = \langle L_{\bar{1}}^{2}, L_{\bar{1}}^{2} \rangle = \{0\} \quad , \quad \langle L_{\bar{1}}^{1}, L_{\bar{1}}^{2} \rangle = L_{\bar{0}} \ . \tag{2.27}$$

Proof

To begin with we derive the following special case of proposition 3 .

Lemma 5

Let L be a simple Lie superalgebra. Suppose that $L_{\bar{1}}$ is the direct sum

$$L_{\bar{1}} = \bigoplus_{s=1}^{r} L_{\bar{1}}^{s} \tag{2.28}$$

of non-zero $L_{\bar{0}}$ - invariant subspaces $L_{\bar{1}}^{s}$, $1 \leq s \leq r$; $r \geq 1$.

Then $r = 1$ or $r = 2$ and in the case $r = 2$ the equations (2.27) are valid.

Proof of the lemma

In the case $r = 1$ there is nothing to prove. Let us consider the case $r = 2$; once this has been established the rest will follow immediately.

We shall prove that

$$J = \langle L_{\bar{1}}^{1}, L_{\bar{1}}^{1} \rangle \oplus \langle L_{\bar{1}}^{1}, \langle L_{\bar{1}}^{1}, L_{\bar{1}}^{1} \rangle\rangle \tag{2.29}$$

is an ideal of L .

The $L_{\bar{0}}$ - invariance of J is obvious, furthermore, we see at once that

$$\langle L_{\bar{1}}^{1}, J \rangle \subset J \ . \tag{2.30}$$

We next remark that

$$\langle\langle L_{\bar{1}}^{1}, L_{\bar{1}}^{1} \rangle, L_{\bar{1}}^{2} \rangle \subset \langle L_{\bar{1}}^{1}, \langle L_{\bar{1}}^{1}, L_{\bar{1}}^{2} \rangle\rangle \subset L_{\bar{1}}^{1} \ . \tag{2.31}$$

Since $L_{\bar{1}}^{2}$ is $L_{\bar{0}}$ - invariant it follows that

$$\langle L_{\bar{1}}^{2}, \langle L_{\bar{1}}^{1}, L_{\bar{1}}^{1} \rangle\rangle = \{0\} \ . \tag{2.32}$$

This implies

$$\langle L_{\bar{1}}^2, \langle L_{\bar{1}}^1, \langle L_{\bar{1}}^1, L_{\bar{1}}^1 \rangle\rangle\rangle \subset \langle\langle L_{\bar{1}}^2, L_{\bar{1}}^1 \rangle, \langle L_{\bar{1}}^1, L_{\bar{1}}^1 \rangle\rangle \subset \langle L_{\bar{1}}^1, L_{\bar{1}}^1 \rangle . \quad (2.33)$$

Combining the equations (2.32) and (2.33) we obtain

$$\langle L_{\bar{1}}^2, J \rangle \subset J . \quad (2.34)$$

Thus we have shown that J is an ideal of L . Since $J \neq L$ we conclude that

$$\langle L_{\bar{1}}^1, L_{\bar{1}}^1 \rangle = \{0\} \quad (2.35)$$

and similarly

$$\langle L_{\bar{1}}^2, L_{\bar{1}}^2 \rangle = \{0\} . \quad (2.36)$$

But then part 2) of lemma 2 yields

$$\langle L_{\bar{1}}^1, L_{\bar{1}}^2 \rangle = \langle L_{\bar{1}}, L_{\bar{1}} \rangle = L_{\bar{0}} . \quad (2.37)$$

Suppose now that $r \geqslant 3$. If $s \in \{1,\dots,r\}$ we have

$$L_{\bar{1}} = L_{\bar{1}}^s \oplus \bigoplus_{i \neq s} L_{\bar{1}}^i \quad (2.38)$$

and the case $r = 2$ implies

$$\langle \bigoplus_{i \neq s} L_{\bar{1}}^i , \bigoplus_{j \neq s} L_{\bar{1}}^j \rangle = \{0\} . \quad (2.39)$$

It follows that

$$\langle L_{\bar{1}}^i, L_{\bar{1}}^j \rangle = \{0\} \quad \text{for all } i, j \in \{1,\dots,r\} \quad (2.40)$$

which is to say that

$$\langle L_{\bar{1}}, L_{\bar{1}} \rangle = \{0\} . \quad (2.41)$$

Since by assumption $L_{\bar{1}} \neq \{0\}$, this is in contradiction to the simplicity of L (see lemma 2).

Having established lemma 5 we shall now prove proposition 3 . Let $L_{\bar{1}}^1$ and $L_{\bar{1}}^2$ be the subspaces of $L_{\bar{1}}$ mentioned therein. We define two sequences $(V_n^i)_{n \geqslant -1}$; $i = 1, 2$, of subspaces of L , as follows. Set

$$V^i_{-1} = L_{\bar{1}} \quad , \quad V^i_0 = L_{\bar{0}} \quad , \quad V^i_1 = L^i_{\bar{1}} \tag{2.42,a}$$

and define inductively

$$V^i_n = \langle L^i_{\bar{1}} , V^i_{n-1} \rangle \quad \text{if } n \geqslant 2 \ . \tag{2.42,b}$$

Note that the subspaces V^i_n are even (resp. odd) if the integer n is even (resp. odd). Furthermore, it is easy to see that for all integers $n \geqslant -1$

$$V^i_n \text{ is } L_{\bar{0}}\text{- invariant} \tag{2.43}$$

$$\langle L_{\bar{1}} , V^i_{n+1} \rangle \subset V^i_n \tag{2.44}$$

$$V^i_{n+2} \subset V^i_n \ . \tag{2.45}$$

From (2.45) we deduce that there exists an integer $m \geqslant 1$ such that

$$V^i_{2m+2} = V^i_{2m} \ . \tag{2.46}$$

Using (2.43) and (2.44) we conclude that $V^i_{2m} \oplus V^i_{2m+1}$ is a graded ideal of L which, obviously, must be equal to $\{0\}$.

Thus we have shown that

$$V^i_n = \{0\} \tag{2.47}$$

if the positive integer n is sufficiently large.

Now let $r \geqslant 0$ be any positive integer. We define

$$J^r_{\bar{0}} = \sum_{s=0}^{r} (V^1_{2(r-s)} \cap V^2_{2s}) \tag{2.48,a}$$

$$J^r_{\bar{1}} = \sum_{s=0}^{r+1} (V^1_{2(r-s)+1} \cap V^2_{2s-1}) \tag{2.48,b}$$

$$J^r = J^r_{\bar{0}} \oplus J^r_{\bar{1}} \ . \tag{2.48,c}$$

Using (2.43) , (2.44) and (2.25) it is easy to see that J^r is a graded ideal of L , for every $r \geqslant 0$. We remark that

$$J^0_{\bar{0}} = L_{\bar{0}} \quad , \quad J^0_{\bar{1}} = L^1_{\bar{1}} + L^2_{\bar{1}} = L_{\bar{1}} \ . \tag{2.49}$$

Evidently,

$$J_{\bar{1}}^{r} \subset (V_{2r+1}^{1} + L_{\bar{1}}^{2}) \cap (L_{\bar{1}}^{1} + V_{2r+1}^{2}) \quad . \tag{2.50}$$

Therefore, if $V_{2r+1}^{1} = \{0\}$ or $V_{2r+1}^{2} = \{0\}$ (which will be the case if r is sufficiently large) then $J_{\bar{1}}^{r} \neq L_{\bar{1}}$ and hence $J^{r} = \{0\}$.

Now let R be the smallest of all integers $r \geq 1$ such that

$$V_{2(r-s)+1}^{1} \cap V_{2s-1}^{2} = \{0\} \quad \text{if } 1 \leq s \leq r \tag{2.51}$$

(see (2.48,b)). Then on the one hand V_{2R-1}^{1} and V_{2R-1}^{2} are different from $\{0\}$, since otherwise $R \geq 2$ and $J^{R-1} = \{0\}$, hence R would not be minimal. In particular, we conclude that $J^{R-1} = L$ and, consequently, $J_{\bar{1}}^{R-1} = L_{\bar{1}}$.
But on the other hand it follows from (2.51) that the sum defining $J_{\bar{1}}^{R-1}$ is direct. Since we already know that the two terms V_{2R-1}^{1} and V_{2R-1}^{2} of this sum are different from {0} we deduce from lemma 5 that all the remaining terms must be equal to {0}. This implies that R = 1, for otherwise R would not be minimal. Thus we have shown that $L_{\bar{1}}$ is the direct sum of $V_{1}^{1} = L_{\bar{1}}^{1}$ and $V_{1}^{2} = L_{\bar{1}}^{2}$.

It is now easy to see that $L_{\bar{1}}^{1}$ and $L_{\bar{1}}^{2}$ are irreducible $L_{\bar{0}}$-modules. For suppose for instance that $\tilde{L}_{\bar{1}}^{1}$ is a proper $L_{\bar{0}}$-invariant subspace of $L_{\bar{1}}^{1}$. Then we can apply the above result to $L_{\bar{1}}^{1}$ and $\tilde{L}_{\bar{1}}^{1} \oplus L_{\bar{1}}^{2}$ and conclude that the sum of these two spaces must be direct, i.e. that $\tilde{L}_{\bar{1}}^{1} = \{0\}$.

Remarks

4) Using the notation introduced in proposition 3 let us define

$$G_{-1} = L_{\bar{1}}^{1} \quad , \quad G_{0} = L_{\bar{0}} \quad , \quad G_{1} = L_{\bar{1}}^{2} \quad . \tag{2.52}$$

Then $(G_n)_{-1 \leq n \leq 1}$ is a Z-gradation of the Lie superalgebra L which is consistent with the Z_2-gradation.

5) Let L be a simple Lie superalgebra. Then there exist only the following two possibilities.

a) The $L_{\bar{0}}$-module $L_{\bar{1}}$ is completely reducible. Then $L_{\bar{1}}$ decomposes into at most two irreducible components.

b) The $L_{\bar{0}}$ - module $L_{\bar{1}}$ is not completely reducible. In this case there exists a (unique) proper $L_{\bar{0}}$ - submodule of $L_{\bar{1}}$ which contains all proper $L_{\bar{0}}$ - submodules of $L_{\bar{1}}$.

In the subsequent paragraphs we shall see that these two possibilities really do occur.

Definition 2

A simple Lie superalgebra L is called *classical* if the $L_{\bar{0}}$ - module $L_{\bar{1}}$ is completely reducible.

Theorem 1

A simple Lie superalgebra L is classical if and only if its Lie algebra $L_{\bar{0}}$ is reductive.

Remark 6)

Whereas one half of the theorem follows directly from standard Lie algebra theory the proof of the other half is rather lengthy. Since we shall not use this characterization to obtain the classification of simple Lie superalgebras we might read off this result from the final classification theorem. However, we think that it is worth-while to derive theorem 1 in a way which is independent of the classification.

Proof of theorem 1

We may assume that $L_{\bar{1}} \neq \{0\}$, for otherwise $L = L_{\bar{0}}$ is a simple Lie algebra. It follows that the adjoint representation ad' of $L_{\bar{0}}$ in $L_{\bar{1}}$ is faithful.

Suppose now that L is classical. Then $L_{\bar{0}}$ is reductive for it has a (finite-dimensional) faithful completely reducible representation.

Conversely, let us assume that $L_{\bar{0}}$ is reductive. We have to show that ad' is completely reducible.

To begin with we reduce the proof to the case where the field K is algebraically closed. In fact, let E be any field containing K and let

$\hat{L} = E \underset{K}{\otimes} L$ be the Lie superalgebra obtained from L by extension of the base field from K to E. Applying proposition 1 of chapter I, §3, n°2 to the adjoint representation of L it is easy to see that $\hat{L}$ is the direct sum of simple graded ideals

$$\hat{L} = \bigoplus_{r=1}^{t} \hat{L}^r . \tag{2.53}$$

It is well-known that the Lie algebra $\hat{L}_{\bar{0}}$ as well as its ideals $\hat{L}^r_{\bar{0}}$ are reductive.

Suppose we can show that for every r the $\hat{L}^r_{\bar{0}}$-module $\hat{L}^r_{\bar{1}}$ is completely reducible. Then the $\hat{L}_{\bar{0}}$-module $\hat{L}_{\bar{1}}$ is also completely reducible and it follows from standard representation theory that the $L_{\bar{0}}$-module $L_{\bar{1}}$ is completely reducible, too.

Thus in the following we may assume that the field K is algebraically closed. By assumption $L_{\bar{0}}$ is the direct product of the semi-simple Lie algebra $L^s_{\bar{0}} = \langle L_{\bar{0}}, L_{\bar{0}} \rangle$ with the center $L^a_{\bar{0}}$ of $L_{\bar{0}}$, i.e.

$$L_{\bar{0}} = L^s_{\bar{0}} \times L^a_{\bar{0}} . \tag{2.54}$$

In view of proposition 3 it is sufficient to prove the following statement:

Suppose that the $L_{\bar{0}}$-module $L_{\bar{1}}$ is not the sum of two proper $L_{\bar{0}}$-submodules. Then $L^a_{\bar{0}} = \{0\}$ and hence the $L_{\bar{0}}$-module $L_{\bar{1}}$ is irreducible.

According to our assumptions there exist a faithful irreducible $L^s_{\bar{0}}$-module V and a faithful $L^a_{\bar{0}}$-module W such that the $L_{\bar{0}}$-module $L_{\bar{1}}$ is isomorphic to the $L^s_{\bar{0}} \times L^a_{\bar{0}}$-module $V \otimes W$. In the following we shall identify $L_{\bar{1}}$ with $V \otimes W$.

Of course, the $L^a_{\bar{0}}$-module W is not the sum of two proper $L^a_{\bar{0}}$-submodules. This implies:

a) There exists a (unique) proper $L^a_{\bar{0}}$-submodule W' of W which contains all proper $L^a_{\bar{0}}$-submodules of W.

b) If $a \in W$ but $a \notin W'$ then a is a cyclic vector for the $L^a_{\bar{0}}$-module W.

c) The elements of $L^a_{\bar{0}}$ are represented in W by nilpotent linear mappings.

In fact, let $S \in L^a_{\bar{0}}$ and let S_W be the homothety of W defined by S. Then

S_W has only one eigenvalue, for otherwise W would decompose into the direct sum of (at least) two non-zero $L^a_{\bar{0}}$ - submodules.

On the other hand we know from lemma 2 that

$$\operatorname{str}(\operatorname{ad}_L S) = 0 . \qquad (2.55)$$

Since the restriction of $\operatorname{ad}_L S$ to $L_{\bar{0}}$ is equal to zero this equation implies

$$\operatorname{Tr}(S_W) = 0 . \qquad (2.56)$$

Thus the sole eigenvalue of S_W is equal to zero.

d) $$S \cdot W \subset W' \text{ for all } S \in L^a_{\bar{0}} . \qquad (2.57)$$

This follows from a) and c).

e) The codimension of W' in W is equal to one.

Let $a \in W$ but $a \notin W'$. According to d) the subspace $Ka + W'$ of W is $L^a_{\bar{0}}$-invariant. We conclude from a) that

$$Ka \oplus W' = W . \qquad (2.58)$$

f) $$L^s_{\bar{0}} \neq \{0\} , \quad \dim V \geq 2 . \qquad (2.59)$$

In fact, if $L^s_{\bar{0}} = \{0\}$, then the relation d) implies that $\langle L_{\bar{0}}, L_{\bar{1}} \rangle \neq L_{\bar{1}}$. In view of lemma 2 this is a contradiction.

After these preliminaries we are now ready to prove our statement: We assume that

$$L^a_{\bar{0}} \neq \{0\} \qquad (2.60)$$

and derive a contradiction.

Let

$$P : L_{\bar{1}} \times L_{\bar{1}} \longrightarrow L^s_{\bar{0}} , \quad B : L_{\bar{1}} \times L_{\bar{1}} \longrightarrow L^a_{\bar{0}} \qquad (2.61)$$

be the bilinear mappings defined by the multiplication in L, i.e.

$$\langle X, Y \rangle = P(X,Y) + B(X,Y) \text{ for all } X, Y \in L_{\bar{1}} . \qquad (2.62)$$

We shall first discuss the mapping B. Lemma 2 shows that $\langle L_{\bar{1}}, L_{\bar{1}} \rangle = L_{\bar{0}}$. Since $L^a_{\bar{0}} \neq \{0\}$ it follows that $B \neq 0$; moreover, B is $L^s_{\bar{0}}$- invariant. This implies that there exists a non-zero $L^s_{\bar{0}}$ - invariant bilinear form ψ on V.

Since the $L_{\bar{0}}^{s}$ - module V is irreducible, ψ is non-degenerate; moreover, any other $L_{\bar{0}}^{s}$ - invariant bilinear form on V is proportional to ψ. We conclude that

$$B(x \otimes a, y \otimes b) = \psi(x,y) A(a,b) \qquad (2.63)$$

for all $x, y \in V$ and $a, b \in W$

with some bilinear mapping

$$A : W \times W \longrightarrow L_{\bar{0}}^{a} . \qquad (2.64)$$

Of course, $A \neq 0$.

It is well-known that ψ is either symmetric or skew-symmetric. Since B is symmetric we see that ψ and A are either both symmetric or else both skew-symmetric. We shall consider these two cases separately.

I) ψ and A are symmetric.

Let $x \in V$ and $a \in W$, $a \notin W'$. Then the Jacobi identity for the three odd elements which are all equal to $x \otimes a$ reads

$$(P(x \otimes a, x \otimes a)x) \otimes a + \psi(x,x) x \otimes A(a,a)a = 0 . \qquad (2.65)$$

Since ψ is symmetric and non-degenerate there exist elements $x \in V$ such that $\psi(x,x) \neq 0$. On the other hand we know that $a \notin W'$ but $A(a,a)a \in W'$ (see d)). It follows that

$$A(a,a)a = 0 \text{ for all } a \in W, a \notin W' . \qquad (2.66)$$

This implies that

$$A(a,a) S_1 \ldots S_r a = 0 \text{ for all } S_1, \ldots, S_r \in L_{\bar{0}}^{a} \text{ and all } r \geq 0 . \qquad (2.67)$$

But a is a cyclic vector for the $L_{\bar{0}}^{a}$ - module W (see b)); hence we conclude that

$$A(a,a) = 0 \text{ for all } a \in W, a \notin W' . \qquad (2.68)$$

Since A is symmetric this implies $A = 0$, a contradiction.

II) ψ and A are skew-symmetric.

This case is more difficult. To begin with let us prove that there exist elements $\bar{a} \in W$, $\bar{a} \notin W'$ and $\bar{b} \in W'$ such that

$$A(\bar{b},\bar{a})\bar{a} \neq 0 . \qquad (2.69)$$

Suppose the contrary; then we would have

$$A(b,a)\,a = 0 \quad \text{for all } a \in W,\ a \notin W' \text{ and } b \in W' . \tag{2.70}$$

In the same way as (2.68) has been derived from (2.66) we could then conclude that

$$A(b,a) = 0 \quad \text{for all } a \in W,\ a \notin W' \text{ and } b \in W' . \tag{2.71}$$

Using the fact that the codimension of W' in W is equal to one (see e)) as well as the skew-symmetry of A it is easy to see that (2.71) leads to the contradiction $A = 0$.

Now let x_1, x_2, x_3 be arbitrary elements of V; we exploit the Jacobi identity for the three odd elements $x_1 \otimes \bar{a}$, $x_2 \otimes \bar{a}$ and $x_3 \otimes \bar{b}$. Using the skew-symmetry of A, the assumption $\bar{a} \notin W'$ and the relations (2.57) and (2.69) this identity implies

$$P(x_1 \otimes \bar{a}, x_2 \otimes \bar{a})\,x_3 \otimes \bar{b} = \{\psi(x_2,x_3)\,x_1 - \psi(x_3,x_1)\,x_2\} \otimes A(\bar{b},\bar{a})\,\bar{a} . \tag{2.72}$$

Since ψ is non-degenerate the curly bracket on the right hand side is not identically zero. Hence there exists a non-zero constant $\sigma \in K$ such that

$$A(\bar{b},\bar{a})\,\bar{a} = \sigma \bar{b} . \tag{2.73}$$

But then equation (2.72) yields

$$P(x_1 \otimes \bar{a}, x_2 \otimes \bar{a})\,x_3 = \sigma\{\psi(x_2,x_3)\,x_1 - \psi(x_3,x_1)\,x_2\} \tag{2.74}$$

$$\text{for all } x_1, x_2, x_3 \in V .$$

If x_1, x_2 run through all elements of V the linear mappings

$$x_3 \longrightarrow \psi(x_2,x_3)\,x_1 - \psi(x_3,x_1)\,x_2 \quad ; \quad x_3 \in V \tag{2.75}$$

span the symplectic Lie algebra $sp(\psi)$ (considered as a vector space). On the other hand, ψ is $L^s_{\bar{0}}$-invariant. Since the representation of $L^s_{\bar{0}}$ in V is faithful we conclude that this representation is an isomorphism of the Lie algebra $L^s_{\bar{0}}$ onto the symplectic Lie algebra $sp(\psi)$.

But then it is well-known that up to a scalar factor there exists a unique $L^s_{\bar{0}}$-invariant bilinear mapping

$$Q : V \times V \longrightarrow L^s_{\bar{0}} ; \tag{2.76,a}$$

in fact, with a suitable normalization of Q we have

$$Q(x_1,x_2)\,x_3 = \psi(x_2,x_3)\,x_1 - \psi(x_3,x_1)\,x_2 \qquad (2.76,b)$$

$$\text{for all } x_1, x_2, x_3 \in V .$$

Now, for any two elements $a_1, a_2 \in W$, the bilinear mapping

$$(x_1,x_2) \longrightarrow P(x_1 \otimes a_1, x_2 \otimes a_2) \quad ; \quad x_1, x_2 \in V \qquad (2.77)$$

is $L^S_{\bar{0}}$ - invariant. We conclude that there exists a bilinear form β on W such that

$$P(x_1 \otimes a_1, x_2 \otimes a_2) = \beta(a_1,a_2)\,Q(x_1,x_2) \qquad (2.78)$$

$$\text{for all } x_1, x_2 \in V \text{ and } a_1, a_2 \in W .$$

Since P and Q are symmetric it follows that β is symmetric, too.

Combining our results we have shown that

$$\langle x_1 \otimes a_1, x_2 \otimes a_2 \rangle = \beta(a_1,a_2)\,Q(x_1,x_2) + \psi(x_1,x_2)\,A(a_1,a_2) \qquad (2.79)$$

$$\text{for all } x_1, x_2 \in V \text{ and } a_1, a_2 \in W .$$

Now let $x_i \in V$ and $a_i \in W$; $i = 1,2,3$. Then the Jacobi identity for the three odd elements $x_i \otimes a_i$ implies that the trilinear mapping

$$\hat{A} : W \times W \times W \longrightarrow W \qquad (2.80,a)$$

which is defined by

$$\hat{A}(a_1,a_2,a_3) = A(a_1,a_2)\,a_3 - \{\beta(a_2,a_3)\,a_1 - \beta(a_3,a_1)\,a_2\} \qquad (2.80,b)$$

$$\text{for all } a_1, a_2, a_3 \in W$$

is totally skew-symmetric. (In the case $\dim V \geq 3$ we can even conclude that $\hat{A} = 0$.)

It follows that

$$A(b,a)\,a = \beta(a,a)\,b - \beta(a,b)\,a \quad \text{for all } a, b \in W . \qquad (2.81)$$

But $A(b,a)\,a$ is an element of W' for all $a, b \in W$ (see d)). Hence we derive from (2.81) that

$$\beta(a,b) = 0 \quad \text{if } a \in W,\ a \notin W' \text{ and } b \in W' \qquad (2.82)$$

and this implies

$$\beta(a,b) = 0 \quad \text{for all } a \in W\,,\ b \in W'\ . \tag{2.83}$$

In view of equation (2.79) we conclude that

$$\langle V \otimes W\,, V \otimes W' \rangle \subset L^a_{\bar{0}}\ . \tag{2.84}$$

But then $L^a_{\bar{0}} \oplus (V \otimes W')$ is an ideal of L (recall the relation (2.57)) which is neither equal to $\{0\}$ nor equal to L. Thus we have arrived at a contradiction and the theorem is proved.

The following corollary is easily derived from the results proved in this section; we shall also include a proof which is independent of theorem 1.

Corollary

We suppose that the field K is algebraically closed.

Let L be a classical simple Lie superalgebra such that the center $L^a_{\bar{0}}$ of $L_{\bar{0}}$ is non-trivial. Then

$$\dim L^a_{\bar{0}} = 1 \tag{2.85}$$

and the $L_{\bar{0}}$ - module $L_{\bar{1}}$ decomposes into the direct sum of two irreducible $L_{\bar{0}}$ - submodules :

$$L_{\bar{1}} = L^1_{\bar{1}} \oplus L^2_{\bar{1}}\ . \tag{2.86}$$

Furthermore, there exists a unique element $C \in L^a_{\bar{0}}$ such that

$$\langle C\,, X \rangle = (-1)^r X \quad \text{for all } X \in L^r_{\bar{1}}\ ;\ r = 1\,,2\ . \tag{2.87}$$

1. Proof

Evidently $L_{\bar{1}} \neq \{0\}$. We have seen in the proof of theorem 1 that the $L_{\bar{0}}$ - module $L_{\bar{1}}$ cannot be irreducible. In view of remark 4) and lemma 4 our corollary is a special case of §1, proposition 3.

2. Proof

Suppose that the $L_{\bar{0}}$ - module $L_{\bar{1}}$ is irreducible. If $A \in L^a_{\bar{0}}$ there exists an element $\alpha \in K$ such that

$$\langle A\,, X \rangle = \alpha X \quad \text{for all } X \in L_{\bar{1}}\ . \tag{2.88}$$

This implies

$$2\alpha L_{\bar{0}} = 2\alpha \langle L_{\bar{1}}, L_{\bar{1}} \rangle = \langle A, \langle L_{\bar{1}}, L_{\bar{1}} \rangle\rangle = \{0\} \tag{2.89}$$

(see lemma 2) and hence $\alpha = 0$. Since the representation of $L_{\bar{0}}$ in $L_{\bar{1}}$ is faithful (see lemma 2) we have shown that $L_{\bar{0}}^{a} = \{0\}$, contrary to our assumption.

The $L_{\bar{0}}$ - module $L_{\bar{1}}$ is completely reducible. Hence we conclude from proposition 3 that the $L_{\bar{0}}$ - module $L_{\bar{1}}$ is the direct sum of two irreducible $L_{\bar{0}}$ - submodules :

$$L_{\bar{1}} = L_{\bar{1}}^{1} \oplus L_{\bar{1}}^{2} \; . \tag{2.90}$$

Let A be any element of $L_{\bar{0}}^{a}$. Then there exist two elements α_r ; $r = 1, 2$, of K such that

$$\langle A, X_r \rangle = \alpha_r X_r \quad \text{for all } X_r \in L_{\bar{1}}^{r} ;\; r = 1, 2 \; . \tag{2.91}$$

It follows that

$$(\alpha_1 + \alpha_2) L_{\bar{0}} = (\alpha_1 + \alpha_2) \langle L_{\bar{1}}^{1}, L_{\bar{1}}^{2} \rangle = \langle A, \langle L_{\bar{1}}^{1}, L_{\bar{1}}^{2} \rangle\rangle = \{0\} \tag{2.92}$$

(see proposition 3) and hence $\alpha_1 + \alpha_2 = 0$. On the other hand the representation of $L_{\bar{0}}$ in $L_{\bar{1}}$ is faithful (see lemma 2). Therefore, our result implies that $\dim L_{\bar{0}}^{a} \leqslant 1$.

3. Cartan subalgebras of a Lie superalgebra

In this section we suppose that the field K is *algebraically closed.*

Let L be a Lie superalgebra and let h be a Cartan subalgebra of the Lie algebra $L_{\bar{0}}$. Then we can construct the weight space decomposition of L with respect to h [12].

Let λ be any linear form on h and let $L^{\lambda}(h)$ be the set of all elements $X \in L$ such that for every element $H \in h$

$$(\mathrm{ad}_L H - \lambda(H))^n (X) = 0 \tag{2.93}$$

provided that the positive integer n (which may depend on H) is sufficiently large.

Evidently, $L^{\lambda}(h)$ is a Z_2 - graded subspace of L . It is well-known that

$$L = \bigoplus_{\lambda \in h^*} L^{\lambda}(h) \tag{2.94}$$

and that

$$\langle L^{\lambda}(h), L^{\mu}(h) \rangle \subset L^{\lambda+\mu}(h) \quad \text{for all } \lambda, \mu \in h^* . \tag{2.95}$$

Let us define

$$\Delta_{\bar{0}} = \{ \lambda \in h^* \mid \lambda \neq 0 , L_{\bar{0}}^{\lambda}(h) \neq \{0\} \} \tag{2.96,a}$$

$$\Delta_{\bar{1}} = \{ \lambda \in h^* \mid L_{\bar{1}}^{\lambda}(h) \neq \{0\} \} \tag{2.96,b}$$

$$\Delta = \Delta_{\bar{0}} \cup \Delta_{\bar{1}} . \tag{2.96,c}$$

The elements of Δ are called the *roots of* L *with respect to* h , more precisely, a root is called *even* (resp. *odd*) if it is an element of $\Delta_{\bar{0}}$ (resp. of $\Delta_{\bar{1}}$). Note that a root may be even *and* odd, note, furthermore, that the linear form 0 on h is (by definition) not an even root of L , however, 0 may be an odd root.

Since h is a Cartan subalgebra of $L_{\bar{0}}$ we have

$$L_{\bar{0}}^{0}(h) = h . \tag{2.97}$$

Thus the equation (2.94) may be written in the form

$$L = h \oplus \bigoplus_{\lambda \in \Delta_{\bar{0}}} L_{\bar{0}}^{\lambda}(h) \oplus \bigoplus_{\lambda \in \Delta_{\bar{1}}} L_{\bar{1}}^{\lambda}(h) . \tag{2.98}$$

We shall now show that any two Cartan subalgebras of $L_{\bar{0}}$ are conjugate to each other under an automorphism of the Lie superalgebra L .

Let D be any nilpotent even derivation of the Lie superalgebra L . Then it is obvious that exp(D) is well-defined and is an automorphism of the Lie superalgebra L .

Now let h be a Cartan subalgebra of $L_{\bar{0}}$ and let λ be an even root of L with respect to h . If $X \in L_{\bar{0}}^{\lambda}(h)$ it follows from (2.95) that $ad_L X$ is nilpotent (recall that $\lambda \neq 0$). Let $\bar{E}(h)$ denote the group of automorphisms of the Lie superalgebra L which is generated by the automorphisms of the form $\exp(ad_L X)$ with $X \in L_{\bar{0}}^{\lambda}(h)$, $\lambda \in \Delta_{\bar{0}}$. If u is any automorphism of L we have

$$u\,\bar{E}(h)\,u^{-1} = \bar{E}(u(h)) \ . \qquad (2.99)$$

Every element of $\bar{E}(h)$ induces an automorphism of the Lie algebra $L_{\bar{0}}$; let $E(h)$ be the group of those automorphisms of $L_{\bar{0}}$ which are obtained in this way. Of course, $E(h)$ is nothing but the group of automorphisms of $L_{\bar{0}}$ which is generated by the automorphisms of the form $\exp(\mathrm{ad}_{L_{\bar{0}}} X)$ with $X \in L_{\bar{0}}^{\lambda}(h)$, $\lambda \in \Delta_{\bar{0}}$.

Now let h' be a second Cartan subalgebra of $L_{\bar{0}}$. Then it is well-known [27] that

$$E(h) = E(h') \qquad (2.100)$$

and that there exists an automorphism $u_0 \in E(h)$ such that

$$u_0(h) = h' \ . \qquad (2.101)$$

Let u be an element of $\bar{E}(h)$ whose restriction to $L_{\bar{0}}$ is equal to u_0 . Then

$$u(h) = h' \qquad (2.102)$$

and

$$\bar{E}(h) = u\,\bar{E}(h)\,u^{-1} = \bar{E}(u(h)) = \bar{E}(h') \ . \qquad (2.103)$$

Therefore, exactly as in the case of Lie algebras, *the group* $\bar{E}(h)$ *does not depend on the special choice of the Cartan subalgebra* h *of* $L_{\bar{0}}$; hence we may simplify the notation and write $\bar{E}$ instead of $\bar{E}(h)$. Moreover, we have shown :

Proposition 4

Suppose that the field K is algebraically closed. Then the group $\bar{E}$ operates transitively on the set of all Cartan subalgebras of the Lie algebra $L_{\bar{0}}$.

According to proposition 4 the choice of a special Cartan subalgebra of $L_{\bar{0}}$ does not cause any lack of generality for the study of L , moreover, it is quite reasonable to *call a Cartan subalgebra of* $L_{\bar{0}}$ *also a Cartan subalgebra of the Lie superalgebra* L .

Remarks

7) Normally it will be obvious from the context which Cartan subalgebra

h of $L_{\bar{0}}$ has been chosen. Then we shall write L^{λ} instead of $L^{\lambda}(h)$.

8) In our applications the representation of h in L will be completely reducible. If this is the case we have for all $\lambda \in h^*$

$$L^{\lambda}(h) = \{ X \in L \mid \langle H , X \rangle = \lambda(H) X \text{ for all } H \in h \} . \qquad (2.104)$$

Note that this remark applies if the Lie algebra $L_{\bar{0}}$ is reductive and if, moreover, the representation of $L_{\bar{0}}$ in $L_{\bar{1}}$ is completely reducible.

9) Some detailed results on the roots of a classical simple Lie superalgebra and on the root space decomposition (2.98) will be given in §3, n°2 and in §4, n°6 of this chapter.

§ 3 LIE SUPERALGEBRAS WHOSE KILLING FORM IS NON-DEGENERATE

We remind the reader that all Lie superalgebras are assumed to be *finite-dimensional.*

1. Some basic general results

Let L be a Lie superalgebra and let ϕ be a bilinear form on L . Recall (see chapter I, §3, $n^0 3$, example 4,c)) that ϕ is called invariant if

$$\phi(\langle A , B \rangle , C) = \phi(A ,\langle B , C \rangle) \quad \text{for all } A , B , C \in L \ . \tag{3.1}$$

Important examples of invariant bilinear forms are the Killing form and, more generally, the bilinear forms associated with the finite-dimensional graded L - modules (see chapter I, §3, $n^0 3$, proposition 3 and definition 5). These bilinear forms are even and supersymmetric (see chapter I, §3, $n^0 3$, definition 4). It turns out that the supersymmetry is quite a "normal feature" of invariant bilinear forms on L , for we have

Proposition 1

Let L be a Lie superalgebra such that $\langle L , L \rangle = L$. Then every invariant bilinear form on L is supersymmetric.

Proof

The proof is trivial : Let ϕ be an invariant bilinear form on L and let $A \in L_\alpha$, $B \in L_\beta$, $C \in L_\gamma$; $\alpha , \beta , \gamma \in Z_2$. Then it is easy to check that

$$\psi(A ,\langle B , C \rangle) - (-1)^{\alpha(\beta+\gamma)} \phi(\langle B , C \rangle , A) \ . \tag{3.2}$$

Since $\langle L , L \rangle = L$, our proposition is proved.

In connection with the bilinear forms associated with graded L - modules the following proposition is of interest :

Proposition 2

Let L be a Lie superalgebra and let ϕ be an invariant bilinear form on L which is associated with some finite-dimensional graded L - module. If ϕ is non-degenerate then the Lie algebra $L_{\bar{0}}$ is reductive.

Proof

Suppose that ϕ is associated with the graded L-module V. Let ϕ^{α}, $\alpha \in Z_2$, be the bilinear form on $L_{\bar{0}}$ which is associated with the $L_{\bar{0}}$-module V_{α}. Then we have

$$\phi(P,Q) = \phi^{\bar{0}}(P,Q) - \phi^{\bar{1}}(P,Q) \quad \text{for all } P, Q \in L_{\bar{0}} . \tag{3.3}$$

Set

$$J^{\alpha} = \{ Q \in L_{\bar{0}} \mid \phi^{\alpha}(Q, L_{\bar{0}}) = \{0\} \} \quad ; \quad \alpha \in Z_2 . \tag{3.4}$$

Since ϕ is even and non-degenerate we derive from equation (3.3) that $J^{\bar{0}} \cap J^{\bar{1}} = \{0\}$. But then a standard result from Lie algebra theory [28] says that $L_{\bar{0}}$ must be reductive.

Corollary

Let L be a simple Lie superalgebra whose Killing form is non-degenerate. Then L is classical.

The following two propositions contain some information on the existence of non-degenerate invariant bilinear forms on a Lie superalgebra.

Proposition 3

Let $G = G_{-1} \oplus G_0 \oplus G_1$ be a consistently Z-graded Lie superalgebra such that

$$\langle G_{-1}, G_1 \rangle = G_0 . \tag{3.5}$$

Suppose we are given a G_0-invariant bilinear form ψ on $G_{-1} \times G_1$.

Then there exists a unique extension of ψ to an even supersymmetric G-invariant bilinear form $\tilde{\psi}$ on G such that the subspaces $G_{\pm 1}$ are totally isotropic.

If ψ is non-degenerate and if the G_0-modules $G_{\pm 1}$ are faithful then $\tilde{\psi}$ is also non-degenerate.

Proof

Our conditions on $\tilde{\psi}$ imply that we have to define

$$\tilde{\psi}(G_n, G_n) = \tilde{\psi}(G_0, G_n) = \tilde{\psi}(G_n, G_0) = \{0\} \quad \text{for } n = \pm 1 , \tag{3.6}$$

furthermore,

$$\tilde{\psi}(X,Y) = -\tilde{\psi}(Y,X) = \psi(X,Y) \quad \text{for all } X \in G_{-1} \text{ and } Y \in G_1 . \tag{3.7}$$

It remains to define the restriction of $\tilde{\psi}$ to $G_0 \times G_0$. By assumption every element of G_0 can be written in the form

$$\sum_{i \in J} \langle X_i, Y_i \rangle \quad \text{with } X_i \in G_{-1}, \; Y_i \in G_1 \text{ for all } i \in J \tag{3.8}$$

and with some finite index set J. Let Q be any element of G_0. Since $\tilde{\psi}$ is expected to be G-invariant we must set

$$\tilde{\psi}\Big(\sum_{i \in J} \langle X_i, Y_i \rangle, Q \Big) = \sum_{i \in J} \psi(X_i, \langle Y_i, Q \rangle) . \tag{3.9}$$

Of course, we have to show that this equation really defines a bilinear form on $G_0 \times G_0$. This will be the case if we can prove that the vanishing of the element (3.8) implies the vanishing of the right hand side in equation (3.9)

Because of $\langle G_{-1}, G_1 \rangle = G_0$ we may assume that $Q = \langle X, Y \rangle$ with $X \in G_{-1}$, $Y \in G_1$. But then it is easy to check that

$$\sum_{i \in J} \psi(X_i, \langle Y_i, \langle X, Y \rangle \rangle) = \psi\Big(\Big\langle \sum_{i \in J} \langle X_i, Y_i \rangle, X \Big\rangle, Y \Big) \tag{3.10}$$

and this yields the required result.

It is now straightforward to show that the restriction of $\tilde{\psi}$ to $G_0 \times G_0$ (which has just been defined) is symmetric (which implies that $\tilde{\psi}$ is supersymmetric) and that $\tilde{\psi}$ is G-invariant.

Finally, let us assume that ψ is non-degenerate and that the representations of G_0 in $G_{\pm 1}$ are faithful. To prove that $\tilde{\psi}$ is non-degenerate it is sufficient to show that the restriction of $\tilde{\psi}$ to $G_0 \times G_0$ is non-degenerate.

Let Q be an element of G_0 such that

$$\tilde{\psi}(G_0, Q) = \{0\} . \tag{3.11}$$

Then we have

$$\psi(G_{-1},\langle G_1, Q\rangle) = \tilde{\psi}(\langle G_{-1}, G_1\rangle, Q) = \{0\} . \qquad (3.12)$$

By assumption this implies $\langle G_1, Q\rangle = \{0\}$ and hence $Q = 0$, as required.

Remark 1)

Proposition 3 remains valid even if G is not finite-dimensional.

Proposition 4

Let L be a classical simple Lie superalgebra such that the center of $L_{\bar{0}}$ is non-trivial. Then the Killing form of L is non-degenerate.

Proof

Denote the Killing form of L by ϕ. If the field K is algebraically closed the proposition follows directly from the corollary to theorem 1 in §2 : If C is the element mentioned in this corollary we have

$$\phi(C,C) = -\dim L_{\bar{1}} ; \qquad (3.13)$$

thus $\phi \neq 0$ and hence ϕ is non-degenerate (see §2, proposition 2).

Suppose now that the field K is arbitrary. Let E be an algebraically closed extension field of K and let $\hat{L} = E \otimes_K L$ denote the Lie superalgebra which is obtained from L by extension of the base field from K to E. From the beginning of the proof of theorem 1 in §2 we know that the Lie superalgebra $\hat{L}$ is the direct sum of graded ideals $\hat{L}^r$, $1 \leqslant r \leqslant t$, which are all classical simple Lie superalgebras. Since at least one of the Lie algebras $\hat{L}^r_{\bar{0}}$ has a non-trivial center the first part of our proof, combined with the subsequent lemma, shows that the Killing form of $\hat{L}$ is non-zero. But then the Killing form ϕ of L is also non-zero, hence ϕ is non-degenerate.

Lemma 1

Let L be a Lie superalgebra and let L' be a graded ideal of L . If ϕ (resp. ϕ') is the Killing form of L (resp. L') then the restriction of ϕ to L' is equal to ϕ' .

If L' and L" are two graded ideals of L such that $\langle L', L'' \rangle = \{0\}$ then these ideals are orthogonal with respect to ϕ .

The proof is obvious.

Proposition 5

Let L be a Lie superalgebra whose Killing form is non-degenerate. Then every superderivation of L is inner, i.e. is of the form ad A with $A \in L$ (see chapter I, §1, in particular example 4)).

Proof

Let $\mathfrak{D}(L)$ be the Lie superalgebra of superderivations of L (see loc. cit.) and let $\tilde{\phi}$ be the Killing form of $\mathfrak{D}(L)$. We know that

$$\mathrm{ad} : L \longrightarrow \mathfrak{D}(L) \tag{3.14}$$

is a homomorphism of Lie superalgebras. Since the Killing form of L is non-degenerate this homomorphism is injective. Moreover, the image ad L of L is a graded ideal of $\mathfrak{D}(L)$; in fact, it follows from the very definition of a superderivation that

$$\langle D , \mathrm{ad}\, A \rangle = \mathrm{ad}\, D(A) \quad \text{for all } D \in \mathfrak{D}(L) \text{ and } A \in L \ . \tag{3.15}$$

According to lemma 1 the restriction of $\tilde{\phi}$ to ad L is equal to the Killing form of ad L , thus (by assumption) this restriction is non-degenerate.

On the other hand we conclude from equation (3.15) that

$$\tilde{\phi}(\mathrm{ad}\, D(A) , \mathrm{ad}\, B) = \tilde{\phi}(D , \mathrm{ad}\langle A , B\rangle) \tag{3.16}$$

$$\text{for all } D \in \mathfrak{D}(L) \text{ and } A , B \in L \ .$$

Now let

$$J = \{ D \in \mathfrak{D}(L) \mid \tilde{\phi}(D, \mathrm{ad}\, L) = \{0\} \} \ . \tag{3.17}$$

Then equation (3.16) and the foregoing remark imply that

$$D(A) = 0 \quad \text{for all } D \in J \text{ and } A \in L \ . \tag{3.18}$$

Thus we have shown that $J = \{0\}$ and the proposition is proved.

Theorem 1

Let L be a Lie superalgebra containing no non-zero abelian graded ideals. Suppose that there exists a homogeneous non-degenerate invariant bilinear form ϕ on L .

Then L has only a finite number of minimal graded ideals L^r, $1 \leq r \leq t$, and L is their direct sum. The ideals L^r are simple Lie superalgebras and they are mutually orthogonal with respect to the (necessarily supersymmetric) bilinear form ϕ . Any left or right ideal of L is graded and is equal to $\bigoplus_{r \in R} L^r$ with a suitable subset R of $\{1, \ldots, t\}$.

Proof

The argument is well-known [29]. If J is a graded ideal of L then

$$J^{\perp} = \{ A \in L \mid \phi(A, J) = \{0\} \} \tag{3.19}$$

is also a graded ideal of L .

Now suppose that J is a minimal (but, of course, non-zero) graded ideal of L . Then $J^{\perp} \cap J$ is a graded ideal of L and hence is equal to $\{0\}$ or to J . In the latter case we have $J \subset J^{\perp}$, which implies

$$\phi(L, \langle J, J \rangle) = \phi(\langle L, J \rangle, J) \subset \phi(J^{\perp}, J) = \{0\} . \tag{3.20}$$

Since ϕ is non-degenerate we conclude that $\langle J, J \rangle = \{0\}$. But by assumption L does not contain any non-zero abelian graded ideals; thus we have arrived at a contradiction.

It follows that

$$L = J^{\perp} \oplus J \tag{3.21}$$

and

$$\langle J^{\perp}, J \rangle \subset J^{\perp} \cap J = \{0\} . \tag{3.22}$$

Consequently, any graded ideal of J or $J^{\perp}$ is a graded ideal of L . This shows that J is a simple Lie superalgebra and that $J^{\perp}$ does not contain any non-zero abelian graded ideals. Obviously, the restriction $\phi^{\perp}$ of ϕ to $J^{\perp}$ is non-degenerate. Thus the pair $(J^{\perp}, \phi^{\perp})$ satisfies the same conditions as (L, ϕ) does. Induction with respect to $\dim L$ then implies that

$$L = \bigoplus_{r=1}^{t} L^r \tag{3.23}$$

where the L^r, $1 \leqslant r \leqslant t$, are minimal (and hence simple) graded ideals of L which are mutually orthogonal with respect to ϕ in the sense that

$$\phi(L^r, L^s) = \{0\} \quad \text{if } 1 \leqslant r < s \leqslant t \,. \tag{3.24}$$

Evidently, our result implies that $\langle L, L\rangle = L$. Therefore, the bilinear form ϕ is supersymmetric (see proposition 1) and equation (3.24) may be generalized to read

$$\phi(L^r, L^s) = \{0\} \quad \text{if } r, s \in \{1, \ldots, t\}\,;\; r \neq s \,. \tag{3.25}$$

Now let L' be an (a priori not necessarily graded) left ideal of L. If $s \in \{1, \ldots, t\}$ the intersection $L^s \cap L'$ is a left ideal of L^s, hence $L^s \cap L'$ is equal to $\{0\}$ or to L^s (see §2, proposition 1). In the first case we have

$$\langle L^s, L'\rangle \subset L^s \cap L' = \{0\} \tag{3.26}$$

and hence

$$L' \subset \bigoplus_{r \neq s} L^r \,, \tag{3.27}$$

in the second case it follows that $L^s \subset L'$. Obviously, this implies that

$$L' = \bigoplus_{r \in R} L^r \tag{3.28}$$

with a suitable subset R of $\{1, \ldots, t\}$. In particular, every minimal graded ideal of L is equal to some L^r.

The case of a right ideal L' is treated similarly, hence our theorem is proved.

Corollary

The Killing form of a Lie superalgebra L is non-degenerate if and only if L is the direct product of classical simple Lie superalgebras whose Killing forms are non-degenerate.

If this is the case the Lie algebra $L_{\bar{0}}$ is reductive and the representation of $L_{\bar{0}}$ in $L_{\bar{1}}$ is completely reducible.

Proof

Let L be a Lie superalgebra whose Killing form ϕ is non-degenerate. We want to apply theorem 1 , hence we have to show that L does not contain any non-zero abelian graded ideals.

Let J be an abelian graded ideal of L . Then we have

$$\langle L,\langle J,L\rangle\rangle \subset J \quad , \quad \langle L,\langle J,J\rangle\rangle = \{0\} . \tag{3.29}$$

It is easy to see that this implies

$$\phi(L,J) = \{0\} . \tag{3.30}$$

Since ϕ is non-degenerate it follows that $J = \{0\}$, as required.

Thus we may apply theorem 1 . The simple graded ideals L^r are orthogonal with respect to ϕ , hence the restriction of ϕ to L^r is non-degenerate. But according to lemma 1 the restriction of ϕ to L^r is the Killing form of L^r. In particular it follows that the simple Lie superalgebras L^r are classical (see the corollary to proposition 2). This implies that the Lie algebra $L_{\bar{0}}$ is reductive (see also proposition 2) and that the representation of $L_{\bar{0}}$ in $L_{\bar{1}}$ is completely reducible.

The converse is obvious (see lemma 1).

Remark 2)

The reader will have noticed that the Lie superalgebras whose Killing form is non-degenerate are to some extent similar to the semi-simple Lie algebras. Because of this fact (which is not very surprising, of course) these Lie superalgebras have been called strictly semi-simple. It should be stressed, however, that this similarity is rather superficial, the most important difference being that the finite-dimensional graded representations of a strictly semi-simple Lie superalgebra are not necessarily completely reducible. We shall come back to this point in chapter III, § § 2 and 3 .

2. The root space decomposition of a Lie superalgebra whose Killing form is non-degenerate

In this section we suppose that the field K is *algebraically closed*.

We consider a Lie superalgebra L such that :

a) The Lie algebra $L_{\bar{0}}$ is reductive.

b) The representation of $L_{\bar{0}}$ in $L_{\bar{1}}$ is completely reducible.

c) There exists a non-degenerate even supersymmetric invariant bilinear form ϕ on L .

Note that these assumptions are fulfilled if the Killing form of L is non-degenerate (see the corollary to theorem 1).

Let h be a Cartan subalgebra of the Lie algebra $L_{\bar{0}}$. Using the notation introduced in §2, $n^{o}3$ we shall exploit the existence of the bilinear form ϕ to get additional information on the roots of L as well as the root space decomposition

$$L = \bigoplus_{\lambda \in h^*} L^{\lambda}(h) \quad . \tag{3.31}$$

Only the most elementary results are derived since for our proof of the classification theorem these will be sufficient.

Recall (see §2, $n^{o}3$, remark 8)) that in the present case

$$L^{\lambda}(h) = \{ X \in L \mid \langle H , X \rangle = \lambda(H) X \text{ for all } H \in h \} . \tag{3.32}$$

The $L_{\bar{0}}$ - invariance of ϕ combined with the fact that ϕ is even implies that for all $\lambda , \mu \in h^*$ and all $\alpha , \beta \in Z_2$

$$\phi (L^{\lambda}_{\alpha}(h) , L^{\mu}_{\beta}(h)) = \{0\} \text{ if } \lambda + \mu \neq 0 \text{ or if } \alpha + \beta \neq \bar{0} . \tag{3.33}$$

Since ϕ is non-degenerate it follows that for all $\lambda \in h^*$ and all $\alpha \in Z_2$ the restriction of ϕ to $L^{\lambda}_{\alpha}(h) \times L^{-\lambda}_{\alpha}(h)$ is non-degenerate.

In particular, the restriction of ϕ to $h = L^{0}_{\bar{0}}(h)$ is non-degenerate. Thus we can define as usual a non-degenerate symmetric bilinear form (|) on h^* , as follows.

Let $\lambda \in h^*$ be any linear form on h. Then there exists a unique element $H_\lambda \in h$ such that

$$\lambda(H) = \phi(H_\lambda, H) \quad \text{for all } H \in h . \tag{3.34}$$

If $\lambda, \mu \in h^*$ we define

$$(\lambda|\mu) = \phi(H_\lambda, H_\mu) = \lambda(H_\mu) = \mu(H_\lambda) . \tag{3.35}$$

Now let $\lambda \in h^*$, $\alpha \in Z_2$, and let $X \in L^\lambda_\alpha(h)$, $Y \in L^{-\lambda}_\alpha(h)$. Then we have $\langle X, Y\rangle \in L^0_{\bar{0}}(h) = h$ and the invariance of ϕ implies that

$$\phi(\langle X, Y\rangle, H) = \lambda(H)\,\phi(X,Y) = \phi(\,\phi(X,Y)\,H_\lambda, H) \tag{3.36}$$

$$\text{for all } H \in h .$$

It follows that

$$\langle X, Y\rangle = \phi(X,Y)\,H_\lambda \tag{3.37}$$

$$\text{for all } X \in L^\lambda_\alpha(h),\ Y \in L^{-\lambda}_\alpha(h)\ ;\ \lambda \in h^*,\ \alpha \in Z_2 .$$

We are now ready to derive some properties of the odd roots of L. Let $\lambda \in h^*$ and let $X \in L^\lambda_{\bar{1}}(h)$, $Y \in L^{-\lambda}_{\bar{1}}(h)$. Using the equations (3.37), (3.32) and (3.35) the Jacobi identity for the three elements X, X, Y reads

$$\langle\langle X, X\rangle, Y\rangle = -2\,(\lambda|\lambda)\,\phi(X,Y)\,X . \tag{3.38}$$

Suppose now that λ is an odd root such that $(\lambda|\lambda) \neq 0$. Then the right hand side of equation (3.38) is not identically zero. On the other hand, $\langle X, X\rangle \in L^{2\lambda}_{\bar{0}}(h)$ and the space $L^{2\lambda}_{\bar{0}}(h)$ is at most one-dimensional. This implies :

<u>Lemma</u> 2

Let λ be an odd root of L such that $(\lambda|\lambda) \neq 0$. Then 2λ is an even root of L and

$$\dim L^\lambda_{\bar{1}}(h) = \dim L^{-\lambda}_{\bar{1}}(h) = 1 . \tag{3.39}$$

A slight modification of the above argument yields the following lemma.

<u>Lemma</u> 3

Let λ and μ be two odd roots of L such that $\lambda \neq \pm\mu$. If $(\lambda|\mu) \neq 0$ then $\lambda + \mu$ or $\lambda - \mu$ is an even root of L.

Proof

Choose elements $X \in L_{\bar{1}}^{\lambda}(h)$ and $Y \in L_{\bar{1}}^{-\lambda}(h)$ such that $\phi(X,Y) \neq 0$ and let Z be any non-zero element of $L_{\bar{1}}^{\mu}(h)$. Then the Jacobi identity for X, Y, Z means

$$\langle\langle Z, X\rangle, Y\rangle + \langle\langle Z, Y\rangle, X\rangle = -(\lambda|\mu)\,\phi(X,Y)\,Z . \qquad (3.40)$$

Since the right hand side is non-zero, at least one of the two elements $\langle Z, X\rangle \in L_{\bar{0}}^{\mu+\lambda}(h)$ and $\langle Z, Y\rangle \in L_{\bar{0}}^{\mu-\lambda}(h)$ must be non-zero.

For later reference let us now more explicitly exploit the fact that the Lie algebra $L_{\bar{0}}$ is reductive. This assumption means that $L_{\bar{0}}$ has the form

$$L_{\bar{0}} = L_{\bar{0}}^{0} \times L_{\bar{0}}^{1} \times \ldots\ldots \times L_{\bar{0}}^{r} \qquad (3.41)$$

where $L_{\bar{0}}^{0}$ is an abelian Lie algebra and where the $L_{\bar{0}}^{i}$, $1 \leq i \leq r$, are simple Lie algebras. (The reader should not confuse $L_{\bar{0}}^{0}$ with $L_{\bar{0}}^{0}(h) = h$.)

It is easy to see that the algebras $L_{\bar{0}}^{j}$, $0 \leq j \leq r$, are mutually orthogonal with respect to ϕ. Hence the restriction ϕ_j of ϕ to $L_{\bar{0}}^{j}$, $0 \leq j \leq r$, is non-degenerate. Since the algebras $L_{\bar{0}}^{i}$, $1 \leq i \leq r$, are simple we conclude that ϕ_i, $1 \leq i \leq r$, is a non-zero multiple of the Killing form of $L_{\bar{0}}^{i}$.

It is well-known that the Cartan subalgebra h takes the form

$$h = h^{0} \times h^{1} \times \ldots\ldots \times h^{r} \qquad (3.42)$$

where h^{j} is a Cartan subalgebra of $L_{\bar{0}}^{j}$, $0 \leq j \leq r$ (in particular, we have $h^{0} = L_{\bar{0}}^{0}$).

The restriction of ϕ_j to h^{j}, $0 \leq j \leq r$, defines a non-degenerate symmetric bilinear form $(\ |\ )_j$ on the dual $(h^{j})^*$ of h^{j} in the same way as $(\ |\ )$ has been defined by the restriction of ϕ to h. Now h^* is canonically isomorphic to $(h^{0})^* \times \ldots\ldots \times (h^{r})^*$, an element $\lambda \in h^*$ being identified with the family $(\lambda_j)_{0 \leq j \leq r}$, where λ_j is the restriction of λ to h^j. If $\lambda = (\lambda_j)_{0 \leq j \leq r}$ and $\mu = (\mu_j)_{0 \leq j \leq r}$ are two elements of h^* then

$$(\lambda|\mu) = \sum_{j=0}^{r} (\lambda_j|\mu_j)_j \quad . \tag{3.43}$$

In the following we shall omit the subscript j on $(\ |\ )_j$.

§ 4 THE CLASSICAL SIMPLE LIE SUPERALGEBRAS

In this paragraph we shall introduce several families of classical simple Lie superalgebras over an arbitrary field [3,5,6]. It turns out (see the subsequent paragraph) that over an algebraically closed field no other classical simple Lie superalgebras with a non-vanishing odd subspace do exist.

1. The general linear Lie superalgebra pl(V)

Let $V = V_{\bar{0}} \oplus V_{\bar{1}}$ be a finite-dimensional Z_2 - graded vector space with

$$\dim V_{\bar{0}} = n \quad , \quad \dim V_{\bar{1}} = m \; ; \tag{4.1}$$

in the following, we shall always assume that

$$n \, , m \geq 1 \quad . \tag{4.2}$$

The starting point for all the constructions in this paragraph is the general linear Lie superalgebra pl(V) which has been defined in chapter I, §1, example 3). Recall that $pl(V)_\xi$, $\xi \in Z_2$, consists of the linear mappings X of V into itself which are homogeneous of degree ξ , and that

$$\langle X , Y \rangle = X \circ Y - (-1)^{\xi\eta} Y \circ X \tag{4.3}$$

$$\text{if } X \in pl(V)_\xi \, , \; Y \in pl(V)_\eta \; ; \; \xi , \eta \in Z_2 \; .$$

The Lie superalgebra pl(V) has a natural Z - gradation. In fact, let us introduce a Z - gradation in the vector space V by the requirement that

$$V_0 = V_{\bar{0}} \quad , \quad V_1 = V_{\bar{1}} \quad , \quad V_i = \{0\} \text{ if } i \notin \{0,1\} \; . \tag{4.4}$$

We define for all $j \in Z$

$$pl(V)_j = \{ X \in pl(V) \mid X(V_i) \subset V_{i+j} \text{ for all } i \in Z \} \; . \tag{4.5}$$

Obviously, this defines a Z - gradation of the Lie superalgebra pl(V) which is consistent with the Z_2 - gradation.

Next we describe the algebra pl(V) in a matrix notation. Let $e_1, \ldots , e_n$ be a basis of $V_{\bar{0}}$ and let $e_{n+1}, \ldots , e_{n+m}$ be a basis of $V_{\bar{1}}$. The elements of pl(V) are linear mappings of V into itself, hence they may be characterized by their matrices with respect to the basis $(e_r)_{1 \leqslant r \leqslant n+m}$ of V. If $X \in$ pl(V), the matrix $(X_{sr})_{1 \leqslant s,r \leqslant n+m}$ of X is introduced by

$$X e_r = \sum_{s=1}^{n+m} X_{sr} e_s \quad , \quad 1 \leqslant r \leqslant n+m \; ; \tag{4.6}$$

for simplicity, this matrix will also be denoted by X.

It is advantageous to write these matrices in a block form

$$X = \begin{pmatrix} A & B \\ C & D \end{pmatrix} \tag{4.7}$$

with A an $n \times n$ matrix, B an $n \times m$ matrix,
C an $m \times n$ matrix, D an $m \times m$ matrix.

The "diagonal" block matrices $\begin{pmatrix} A & 0 \\ 0 & D \end{pmatrix}$ belong to the elements of the Lie algebra $pl(V)_0$, the matrices of the form $\begin{pmatrix} 0 & B \\ 0 & 0 \end{pmatrix}$ (resp. $\begin{pmatrix} 0 & 0 \\ C & 0 \end{pmatrix}$) belong to the elements of the subspace $pl(V)_{-1}$ (resp. $pl(V)_1$).

Let

$$X = \begin{pmatrix} A & B \\ C & D \end{pmatrix} \quad , \quad \tilde{X} = \begin{pmatrix} \tilde{A} & \tilde{B} \\ \tilde{C} & \tilde{D} \end{pmatrix} \tag{4.8}$$

be two such matrices; then the multiplication in pl(V) corresponds to

$$\langle X , \tilde{X} \rangle = \begin{pmatrix} A\tilde{A} - \tilde{A}A + B\tilde{C} + \tilde{B}C & B\tilde{D} - \tilde{B}D + A\tilde{B} - \tilde{A}B \\ C\tilde{A} - \tilde{C}A + D\tilde{C} - \tilde{D}C & D\tilde{D} - \tilde{D}D + C\tilde{B} + \tilde{C}B \end{pmatrix} . \tag{4.9}$$

The Z-graded Lie superalgebra of matrices which has just been defined will be denoted by pl(n,m); by definition, pl(n,m) is isomorphic to pl(V). Evidently,

$$pl(n,m)_0 \simeq gl(n) \times gl(m) \tag{4.10}$$

$$\dim pl(n,m)_{\pm 1} = n\,m \, . \tag{4.11}$$

The representations of $pl(n,m)_0$ in $pl(n,m)_{\pm 1}$ are easily read off from equation (4.9).

Remarks

1) Let V' be the Z_2 - graded vector space such that $V'_{\bar{0}} = V_{\bar{1}}$ and $V'_{\bar{1}} = V_{\bar{0}}$. We know that pl(V) = pl(V') (see chapter I, §3, n°1, remark 2)). Using in V' the basis $e_{n+1},\dots,e_{n+m}, e_1,\dots,e_n$ we see that the mapping

$$\begin{pmatrix} A & B \\ C & D \end{pmatrix} \longrightarrow \begin{pmatrix} D & C \\ B & A \end{pmatrix} \tag{4.12}$$

is an isomorphism of the Lie superalgebra pl(n,m) onto the Lie superalgebra pl(m,n) .

2) In chapter I, §3, n°2, C, remark 5) we have noted that the mapping $X \longrightarrow -{}^TX$ is an isomorphism of the Lie superalgebra pl(V) onto the Lie superalgebra pl(V*) (here ${}^TX$ denotes the supertranspose of $X \in$ pl(V)). Let us introduce in V* the basis dual to $(e_r)_{1\leqslant r\leqslant n+m}$. If $X \in$ pl(V) is described by the block matrix $\begin{pmatrix} A & B \\ C & D \end{pmatrix}$, the supertranspose ${}^TX$ is given by the block matrix

$${}^T\begin{pmatrix} A & B \\ C & D \end{pmatrix} = \begin{pmatrix} {}^tA & -{}^tC \\ {}^tB & {}^tD \end{pmatrix} . \tag{4.13}$$

Therefore, the mapping

$$\begin{pmatrix} A & B \\ C & D \end{pmatrix} \longrightarrow - \begin{pmatrix} {}^tA & -{}^tC \\ {}^tB & {}^tD \end{pmatrix} \tag{4.14}$$

is an automorphism of the Lie superalgebra pl(n,m). Note that this automorphism interchanges the subspaces $\text{pl}(n,m)_{\pm 1}$. Thus the Z - graded Lie superalgebra which is obtained from pl(n,m) by inversion of the Z - gradation (see chapter 0, §2, 5)) is isomorphic to the original one.

Finally, let us recall that in chapter I, equation (3.50) we have defined, for any element $X \in$ pl(V) , its supertrace str(X) . Transcribed to the block matrix notation this definition takes the form

$$\text{str}\begin{pmatrix} A & B \\ C & D \end{pmatrix} = \text{Tr}(A) - \text{Tr}(D) . \tag{4.15}$$

Then

$$(X,Y) \longrightarrow \text{str}(XY) \quad ; \quad X , Y \in \text{pl}(n,m) \tag{4.16}$$

is a non-degenerate even supersymmetric invariant bilinear form on the Lie superalgebra pl(n,m). On the other hand the Killing form of pl(n,m) is given by

$$(X,Y) \longrightarrow 2(n-m)\,\mathrm{str}(XY) - 2\,\mathrm{str}(X)\,\mathrm{str}(Y) . \qquad (4.17)$$

The notation introduced in the present section will be used throughout the whole paragraph. The algebras which we are going to construct are either subalgebras of pl(V) resp. pl(n,m) or else quotients of such algebras modulo some one-dimensional ideal. We shall define these algebras in a basis-independent way, however, their detailed properties are described in the matrix notation.

2. The special linear Lie superalgebra spl(V)

In this section we shall discuss the "graded analogue" to the special linear Lie algebra of a vector space. We know (see chapter I, equation (3.51)) that the supertrace is an even invariant linear form on pl(V):

$$\mathrm{str}(\langle X , Y \rangle) = 0 \quad \text{for all } X , Y \in \mathrm{pl}(V) . \qquad (4.18)$$

It follows that

$$\mathrm{spl}(V) = \{ X \in \mathrm{pl}(V) \mid \mathrm{str}(X) = 0 \} \qquad (4.19)$$

is a graded ideal of pl(V) of codimension one; we call spl(V) the *special linear Lie superalgebra of the graded vector space* V. It is easy to see that spl(V) is the commutator algebra of pl(V) (see chapter I, §1, example 1)). Note that the Z-gradation of pl(V) (see equation (4.5)) induces a Z-gradation of the Lie superalgebra spl(V) which is, of course, consistent with the Z_2-gradation.

Let spl(n,m) denote the Z-graded ideal of pl(n,m) which corresponds to spl(V). Evidently, the algebra spl(n,m) consists of the block matrices $\begin{pmatrix} A & B \\ C & D \end{pmatrix}$ such that

$$\mathrm{Tr}(A) = \mathrm{Tr}(D) . \qquad (4.20)$$

It follows that

$$\mathrm{spl}(n,m)_0 \simeq \mathrm{sl}(n) \times \mathrm{sl}(m) \times \mathrm{gl}(1) \qquad (4.21)$$

$$\dim \mathrm{spl}(n,m)_{\pm 1} = n\,m\,. \qquad (4.22)$$

In the case $n = m$ the algebra spl(n,n) contains the $2n \times 2n$ unit matrix I_{2n}. Obviously, $K \cdot I_{2n}$ is a graded ideal of spl(n,n), hence we can construct the consistently Z-graded Lie superalgebra $\mathrm{spl}(n,n)/K \cdot I_{2n}$; its Lie algebra is isomorphic to $\mathrm{sl}(n) \times \mathrm{sl}(n)$.

It is easy to see that:

a) The Lie superalgebra spl(n,m) is simple provided that $n \neq m$.

b) The Lie superalgebra $\mathrm{spl}(n,n)/K \cdot I_{2n}$ is simple provided that $n \geq 2$.

c) The Lie algebra of $\mathrm{spl}(1,1)/K \cdot I_2$ is equal to $\{0\}$.

d) The bilinear form $(X,Y) \longrightarrow \mathrm{str}(XY)$ on spl(n,m) is invariant. In the case $n \neq m$ this form is non-degenerate. On the other hand, if $n = m$, this form induces a non-degenerate even invariant bilinear form on the quotient algebra $\mathrm{spl}(n,n)/K \cdot I_{2n}$, for all $n \geq 1$.

   Finally, the Killing form of spl(n,m) is given by

$$(X,Y) \longrightarrow 2\,(n-m)\,\mathrm{str}(XY)\,. \qquad (4.23)$$

   In particular, the Killing forms of spl(n,n) and $\mathrm{spl}(n,n)/K \cdot I_{2n}$ are equal to zero.

<u>Remark</u> 3)

Evidently, the mapping (4.12) induces an isomorphism of the Lie superalgebra spl(n,m) onto the Lie superalgebra spl(m,n).

Furthermore, the mapping (4.14) induces an automorphism of the Lie superalgebra spl(n,m) and, in the case $n = m$, also of the Lie superalgebra $\mathrm{spl}(n,n)/K \cdot I_{2n}$. Thus the inversion of the Z-gradation of spl(n,m) (resp. of $\mathrm{spl}(n,n)/K \cdot I_{2n}$) leads to a Z-graded Lie superalgebra which is isomorphic to the original one.

3. Subalgebras of pl(V) which leave invariant a homogeneous non-degenerate bilinear form on V

Let β be a homogeneous (i.e. even or odd) non-degenerate bilinear form on V and let $L(V,\beta)$ be the Z_2 - graded subalgebra of $pl(V)$ consisting of those elements which leave invariant the bilinear form β (see chapter I, §3, n°3, example 2)). Recall that an element $X \in pl(V)_{\xi}$, $\xi \in Z_2$, belongs to $L(V,\beta)_{\xi}$ if and only if

$$\beta(Xy,z) + (-1)^{\xi\eta}\beta(y,Xz) = 0 \quad \text{for all } y \in V_{\eta},\ z \in V\ ;\ \eta \in Z_2\ . \tag{4.24}$$

Denote the matrix $(\beta(e_r,e_s))_{1\leqslant r,s\leqslant n+m}$ of β also by β. Then the condition (4.24) is equivalent to the following equation for the matrices :

$$^{T}X\beta + (-1)^{\xi}\beta X = 0 \tag{4.25}$$

(the supertranspose ^{T}X of a block matrix X has been defined in equation (4.13)).

Remark 4)

Let g be an automorphism of the Z_2 - graded vector space V and let β_g be the bilinear form on V which is defined by

$$\beta_g(y,z) = \beta(g^{-1}(y),g^{-1}(z)) \quad \text{for all } y\,,z \in V\ . \tag{4.26}$$

The automorphism g induces an automorphism of the Lie superalgebra pl(V) which, evidently, maps $L(V,\beta)$ onto $L(V,\beta_g)$, i.e. we have

$$g \circ L(V,\beta) \circ g^{-1} = L(V,\beta_g)\ . \tag{4.27}$$

In the following we shall restrict our attention to the cases where β is either supersymmetric or skew-supersymmetric (see chapter I, §3, n°3, definition 4).

A. The orthosymplectic Lie superalgebras

In this section β denotes a non-degenerate *even* bilinear form on V which is either supersymmetric or skew-supersymmetric. In this case we shall write osp(V,β) instead of L(V,β) and call osp(V,β) the *orthosymplectic Lie superalgebra defined by* β .

Let $\begin{pmatrix} G & 0 \\ 0 & H \end{pmatrix}$ be the matrix of β , written in block form; thus G is a regular $n \times n$ matrix and H is a regular $m \times m$ matrix, moreover, one of the matrices G , H is symmetric and the other is skew-symmetric. An element of pl(V) belongs to osp(V,β) if and only if its block matrix $\begin{pmatrix} A & B \\ C & D \end{pmatrix}$ satisfies the conditions

$$ {}^{t}A\,G + G\,A = 0 \tag{4.28,a} $$

$$ {}^{t}B\,G - H\,C = 0 \tag{4.28,b} $$

$$ {}^{t}D\,H + H\,D = 0 \ . \tag{4.28,c} $$

It is not difficult to show that :

a) The orthosymplectic Lie superalgebras osp(V,β) are simple.

b) The bilinear form $(X,Y) \longrightarrow \mathrm{str}(XY)$ on osp(V,β) is non-degenerate and invariant.

c) The Killing form of osp(V,β) is given by

$$ (X,Y) \longrightarrow (n-m-2)\,\mathrm{str}(XY) \tag{4.29,a} $$
if β is supersymmetric

$$ (X,Y) \longrightarrow (n-m+2)\,\mathrm{str}(XY) \tag{4.29,b} $$
if β is skew-supersymmetric .

Note that in special cases the Killing form may be equal to zero.

The two cases in which β is either supersymmetric or else skew-supersymmetric are not independent from each other. In fact, let V' be the Z_2-graded vector space such that $V'_{\bar 0} = V_{\bar 1}$ and $V'_{\bar 1} = V_{\bar 0}$. We define a bilinear form β' on V' by setting

$$ \beta'(y,z) = \beta(z,y) \quad \text{for all } y\,,z \in V \ . \tag{4.30} $$

If β is a supersymmetric (resp. skew-supersymmetric) bilinear form on V then β' is a skew-supersymmetric (resp. supersymmetric) bilinear form on V' and we have

$$ \mathrm{osp}(V,\beta) = \mathrm{osp}(V',\beta') \ . \tag{4.31} $$

Therefore, we now may restrict our attention to the case where β is *supersymmetric*. Then the matrix G is symmetric and the matrix H is skew-symmetric. It follows that the dimension of the vector space $V_{\bar{1}}$ is even, $m = 2r$. Furthermore, by a suitable choice of the basis $e_{n+1}, \ldots, e_{n+m}$ of $V_{\bar{1}}$ we can achieve that H takes the form

$$H = \begin{pmatrix} 0 & I_r \\ -I_r & 0 \end{pmatrix} \tag{4.32}$$

where I_r denotes the $r \times r$ unit matrix.

If the field K is algebraically closed we may also assume that the matrix G takes some standard normal form, for example, we can achieve that

$$G = I_n \tag{4.33}$$

where I_n denotes the $n \times n$ unit matrix.

Quite generally we define osp(n,2r) to be the Z_2-graded subalgebra of pl(n,2r) consisting of those block matrices $\begin{pmatrix} A & B \\ C & D \end{pmatrix}$ which satisfy the conditions (4.28) with H and G being given by (4.32) and (4.33), respectively. Obviously, we have

$$\mathrm{osp}(n,2r)_{\bar{0}} \simeq o(n) \times sp(2r) \tag{4.34}$$

$$\dim \mathrm{osp}(n,2r)_{\bar{1}} = 2nr . \tag{4.35}$$

The relation (4.34) explains the name "orthosymplectic". Let us stress that the cases $n = 1$ and $n = 2$ are not excluded.

Since $o(1) = \{0\}$ we have $\mathrm{osp}(1,2r)_{\bar{0}} \simeq sp(2r)$.

The case $n = 2$ deserves special attention. Since o(2) is the one-dimensional Lie algebra we see that $\mathrm{osp}(2,2r)_{\bar{0}}$ has a one-dimensional center. Therefore, if the field K is algebraically closed (or, more generally, if K contains the square roots of -1) then the remark 4) in §2, n°2 as well as the corollary to theorem 1 in §2 apply. Thus $\mathrm{osp}(2,2r)_{\bar{1}}$ decomposes into the direct sum of two $\mathrm{osp}(2,2r)_{\bar{0}}$-irreducible subspaces; this decomposition leads to a Z-gradation of osp(2,2r) which is consistent with the Z_2-gradation and which is fixed up to an inversion (see chapter 0, §2, 5)). Moreover, it is easy to construct an automorphism of the Lie superalgebra osp(2,2r) which interchanges these two irreducible subspaces. If in the following we shall regard osp(2,2r) as a Z - graded Lie superalgebra we shall always refer to one of these two equivalent Z - gradations.

B. The Lie superalgebras b(n)

In this section β denotes a non-degenerate *odd* bilinear form on V . We assume that β is either supersymmetric or skew-supersymmetric (which is to say that β is either symmetric or skew-symmetric); in the subsequent discussion the upper (resp. lower) sign corresponds to the former (resp. latter) case.

Since β is non-degenerate and odd we conclude that n = m . Furthermore, by choosing the basis $e_1, \ldots, e_{2n}$ appropriately, we can achieve that the matrix of β takes the block form $\begin{pmatrix} 0 & I_n \\ \pm I_n & 0 \end{pmatrix}$, where I_n stands for the $n \times n$ unit matrix. Then an element of pl(V) belongs to L(V,β) if and only if its block matrix $\begin{pmatrix} A & B \\ C & D \end{pmatrix}$ satisfies the conditions

$$^{t}A + D = 0 \tag{4.36,a}$$

$$^{t}B \mp B = 0 \tag{4.36,b}$$

$$^{t}C \pm C = 0 \ . \tag{4.36,c}$$

These equations show that the automorphism (4.12) of the Lie superalgebra pl(n,n) maps the subalgebra corresponding to the upper sign onto the subalgebra corresponding to the lower sign, and vice versa. Consequently, we may restrict our attention to one special choice of the sign, i.e. we may assume that β is *supersymmetric* (upper sign).

An element of L(V,β) belongs to spl(V) if and only if its block matrix $\begin{pmatrix} A & B \\ C & D \end{pmatrix}$ satisfies the condition

$$\mathrm{Tr}(A) = \mathrm{Tr}(D) = 0 \ . \tag{4.37}$$

Thus the equation

$$b(n) = \left\{ \begin{pmatrix} A & B \\ C & -{}^{t}A \end{pmatrix} \middle| \ \mathrm{Tr}(A) = 0 \ , \ {}^{t}B = B \ , \ {}^{t}C = -C \right\} \tag{4.38}$$

(where A , B , C are $n \times n$ matrices) defines a Z_2 - graded subalgebra of pl(n,n). Obviously, the Z - gradation of pl(n,n) induces a Z - gradation of b(n) which is consistent with its Z_2 - gradation. We have

$$b(n)_0 \simeq sl(n) \tag{4.39}$$

$$\dim b(n)_{\pm 1} = \frac{1}{2}n(n \mp 1) \, . \tag{4.40}$$

By inversion of the Z - gradation of b(n) we obtain a new consistently Z - graded Lie superalgebra b'(n) . Evidently, b(n) and b'(n) are not isomorphic as Z - graded Lie superalgebras (but they are identical as Lie superalgebras).

Finally, it is easy to prove the following statements :

a) Assume that $n \geq 3$. Then the Lie superalgebra b(n) is simple and there does not exist any non-zero invariant bilinear form on b(n) .

b) The Z - graded Lie superalgebra b(2) is isomorphic to $sl(2)^{\xi}$ (see §1, n^o3, b)).

4. The (f,d) algebras of Gell - Mann, Michel, Radicati

In this section we shall assume that n = m . Let c be an odd linear mapping of V onto itself such that $c^2 = -\,\mathrm{id}$ and let L(c) be the Z_2 - graded subalgebra of pl(V) consisting of those elements which leave c invariant. An even (resp. odd) element of pl(V) belongs to L(c) if and only if it commutes (resp. anticommutes) with c .

If the basis $e_1, \ldots, e_{2n}$ of V is chosen appropriately the block matrix of the linear mapping c takes the form $\begin{pmatrix} 0 & I_n \\ -I_n & 0 \end{pmatrix}$ (once again, I_n denotes the $n \times n$ unit matrix) and an element of pl(V) lies in L(c) if and only if its block matrix takes the form $\begin{pmatrix} A & B \\ B & A \end{pmatrix}$. Thus

$$L(n) = \left\{ \begin{pmatrix} A & B \\ B & A \end{pmatrix} \,\middle|\, A, B \in gl(n) \right\} \tag{4.41}$$

is a Z_2 - graded subalgebra of spl(n,n) . The commutator algebra of L(n) (see chapter I, §1, example 1)) is easily seen to be equal to

$$d(n) = \left\{ \begin{pmatrix} A & B \\ B & A \end{pmatrix} \,\middle|\, A \in gl(n) \, , \, B \in sl(n) \right\} \, . \tag{4.42}$$

Obviously, $K \cdot I_{2n}$ is a graded ideal of $d(n)$. Thus we may construct the Lie superalgebra $d(n) / K \cdot I_{2n}$; it is isomorphic to the (f,d) algebra of Gell-Mann, Michel, Radicati [30].

The latter algebra is usually defined as follows. As a Z_2 - graded vector space it is equal to $sl(n) \times sl(n)$, the even (resp. odd) subspace being equal to $sl(n) \times \{0\}$ (resp. to $\{0\} \times sl(n)$); moreover, the multiplication is given by

$$\begin{aligned} &\langle (A,B) , (A',B') \rangle \qquad (4.43) \\ &= ([A , A'] + \{B , B'\} - \tfrac{2}{n}\operatorname{Tr}(BB')\, I_n , [A , B'] + [B , A']) \\ &\qquad \text{for all } A , B , A', B' \in sl(n) . \end{aligned}$$

Here [,] denotes the commutator and { , } denotes the anticommutator of two matrices. Obviously, the Lie algebra of this Lie superalgebra is equal to $sl(n)$.

Recall that

$$\{B , B'\} = \operatorname{Tr}(BB')\, I_2 \quad \text{if } B , B' \in sl(2) \ ; \qquad (4.44)$$

thus in the case $n = 2$ the product of two odd elements is equal to zero. On the other hand one can show :

a) The (f,d) algebra is simple provided that $n \geq 3$.

b) For every $n \geq 2$ the prescription

$$((A,B) , (A',B')) \longrightarrow \operatorname{Tr}(AB' + BA') \quad \text{if } A , B , A', B' \in sl(n) \qquad (4.45)$$

defines a non-degenerate odd invariant bilinear form on the (f,d) algebra.

5. Comments on the exceptional classical simple Lie superalgebras

For simplicity we shall assume in this section that the field K is *algebraically closed.*

Besides the classical simple Lie superalgebras which have been discussed in the last sections there exist some additional ones which are called *exceptional.*

In chapter I, §1, example 5) we have introduced the algebras $\Gamma(\sigma_1,\sigma_2,\sigma_3)$ (recall that $\sigma_1, \sigma_2, \sigma_3$ are elements of K such that $\sigma_1+\sigma_2+\sigma_3=0$). The Lie algebra of $\Gamma(\sigma_1,\sigma_2,\sigma_3)$ is equal to $sl(2) \times sl(2) \times sl(2)$ and ad' (the representation of the Lie algebra in the odd subspace) is the tensor product of the 2-dimensional fundamental representations of the three factors $sl(2)$. This implies that

$$\dim \Gamma(\sigma_1,\sigma_2,\sigma_3) = 17 . \tag{4.46}$$

It is easy to see that the Killing form of $\Gamma(\sigma_1,\sigma_2,\sigma_3)$ is equal to zero.

The algebra $\Gamma(\sigma_1,\sigma_2,\sigma_3)$ *is simple if and only if* $\sigma_1, \sigma_2, \sigma_3$ *are all different from zero*. If this is the case there exists a non-degenerate even supersymmetric invariant bilinear form on $\Gamma(\sigma_1,\sigma_2,\sigma_3)$.

Some of the algebras $\Gamma(\sigma_1,\sigma_2,\sigma_3)$ are isomorphic. In fact, let $(\sigma_1,\sigma_2,\sigma_3)$ and $(\sigma'_1,\sigma'_2,\sigma'_3)$ be two triples of elements from K such that $\sigma_1+\sigma_2+\sigma_3 = \sigma'_1+\sigma'_2+\sigma'_3 = 0$. Then $\Gamma(\sigma_1,\sigma_2,\sigma_3)$ and $\Gamma(\sigma'_1,\sigma'_2,\sigma'_3)$ are isomorphic if and only if there exists a non-zero element $\tau \in K$ and a permutation π of the set $\{1,2,3\}$ such that

$$\sigma'_i = \tau \cdot \sigma_{\pi i} \quad \text{for } i = 1,2,3 . \tag{4.47}$$

This result shows that essentially the algebras $\Gamma(\sigma_1,\sigma_2,\sigma_3)$ form a one-parameter family.

Finally, let us briefly describe the two remaining exceptional classical simple Lie superalgebras; they are called Γ_2 and Γ_3.

The Lie algebra of Γ_2 is equal to $sl(2) \times G_2$ and ad' is the tensor product of the 2-dimensional fundamental representation of $sl(2)$ with the 7-dimensional fundamental representation of G_2. Consequently, we have

$$\dim \Gamma_2 = 31 . \tag{4.48}$$

The algebra Γ_2 may be explicitly constructed using the octonions.

Remark 5)

Perhaps it is worth-while to mention that Γ_2 is the only simple Lie superalgebra (with a non-vanishing odd subspace) for which an exceptional simple Lie algebra is relevant.

The Lie algebra of Γ_3 is equal to $sl(2) \times o(7)$ and ad' is the tensor product of the 2-dimensional fundamental representation of sl(2) with the 8-dimensional spin representation of o(7). Thus we conclude that

$$\dim \Gamma_3 = 40 \quad . \tag{4.49}$$

The algebra Γ_3 may be explicitly constructed using Clifford algebra techniques.

For the actual construction of the algebras Γ_2 and Γ_3 we refer the reader to the literature [5, 8]. Let us remark that the Killing forms of Γ_2 and Γ_3 are non-degenerate.

6. The root space decomposition of the classical simple Lie superalgebras

Throughout this section we shall assume that the field K is *algebraically closed.*

In the foregoing sections we have described the following classical simple Lie superalgebras :

$spl(n,m)$ with $n, m \geqslant 1$; $n \neq m$

$spl(n,n) / K \cdot I_{2n}$ with $n \geqslant 2$

$osp(n,2r)$ with $n, r \geqslant 1$

$b(n)$ with $n \geqslant 3$

$d(n) / K \cdot I_{2n}$ with $n \geqslant 3$

$\Gamma(\sigma_1, \sigma_2, \sigma_3)$ with $\sigma_i \in K$, $\sigma_i \neq 0$, $\sigma_1 + \sigma_2 + \sigma_3 = 0$

Γ_2 , Γ_3 .

We have already mentioned that, apart from the simple Lie algebras, no other classical simple Lie superalgebras do exist.

In the following L will stand for one of the algebras listed above. We shall use the notation introduced in §2, n°3 and in §3, n°2. Let h be a Cartan subalgebra of $L_{\bar{0}}$; we are going to derive some properties of the root space decomposition

$$L = h \oplus \bigoplus_{\lambda \in \Delta_{\bar{0}}} L_{\bar{0}}^{\lambda} \oplus \bigoplus_{\lambda \in \Delta_{\bar{1}}} L_{\bar{1}}^{\lambda} \tag{4.50}$$

of L with respect to h. Recall that any two Cartan subalgebras of $L_{\bar{0}}$ are conjugate to each other under an automorphism of the Lie superalgebra L; thus our results will not depend on the special choice of h.

By definition, the set $\Delta_{\bar{0}}$ of even roots is nothing but the set of roots of the reductive Lie algebra $L_{\bar{0}}$, furthermore, the set $\Delta_{\bar{1}}$ of odd roots is just the set of weights of the representation ad' of $L_{\bar{0}}$ in $L_{\bar{1}}$. Since in all cases the Lie algebra $L_{\bar{0}}$ and the representation ad' are known explicitly we may use the standard representation theory of semi-simple Lie algebras to obtain the root system $\Delta = \Delta_{\bar{0}} \cup \Delta_{\bar{1}}$ and the root spaces L^{λ} of L.

The (f,d) algebras $d(n) / K \cdot I_{2n}$; $n \geq 3$, play a special role. For them ad' is equivalent to the adjoint representation of $sl(n)$. This implies that

$$\Delta_{\bar{1}} = \{0\} \cup \Delta_{\bar{0}} \quad \text{if } L = d(n) / K \cdot I_{2n} . \tag{4.51}$$

In the following proposition these algebras will be disregarded.

Proposition 1

The field K is supposed to be algebraically closed. Let L be one of the classical simple Lie superalgebras listed above. We consider the roots and the root space decomposition of L with respect to some Cartan subalgebra h of $L_{\bar{0}}$.

a) If L is not equal to one of the algebras $d(n) / K \cdot I_{2n}$; $n \geq 3$, then

$$0 \notin \Delta_{\bar{1}} \quad \text{and} \quad \Delta_{\bar{0}} \cap \Delta_{\bar{1}} = \phi . \tag{4.52}$$

b) If L is not equal to one of the algebras $spl(2,2) / K \cdot I_4$, $b(4)$ or $d(n) / K \cdot I_{2n}$; $n \geq 3$, then

$$\dim L^{\lambda} = 1 \quad \text{for every } \lambda \in \Delta . \tag{4.53}$$

c) Let us suppose that L is not equal to one of the algebras $b(3)$ or $d(n) / K \cdot I_{2n}$; $n \geq 3$. We consider two roots λ and μ of L which are proportional :

$$\mu = r \lambda \quad \text{with some } r \in K . \tag{4.54}$$

If λ , μ are both even or both odd, then $r = \pm 1$; if λ is odd and μ is even, then $r = \pm 2$.

d) We next suppose that L is not equal to one of the algebras b(n) or $d(n) / K \cdot I_{2n}$ with $n \geq 3$. Then there exists a non-degenerate even supersymmetric invariant bilinear form on L . It follows that

$$- \Delta_{\alpha} = \Delta_{\alpha} \quad \text{for } \alpha \in Z_2 \; . \tag{4.55}$$

e) Suppose that L is not equal to one of the algebras $spl(2,2) / K \cdot I_4$ or b(n) , $d(n) / K \cdot I_{2n}$ with $n \geq 3$. Let $\alpha , \beta \in Z_2$. If $\lambda \in \Delta_{\alpha}$, $\mu \in \Delta_{\beta}$ and $\lambda + \mu \in \Delta_{\alpha+\beta}$, then

$$\langle L^{\lambda}_{\alpha} , L^{\mu}_{\beta} \rangle = L^{\lambda+\mu}_{\alpha+\beta} \; . \tag{4.56}$$

Proof

The statements a), b), c) follow by an inspection of the various cases, statement d) has been proved in the previous sections.

Finally, let us prove e). We already know that

$$\langle L^{\lambda}_{\alpha} , L^{\mu}_{\beta} \rangle \subset L^{\lambda+\mu}_{\alpha+\beta} \tag{4.57}$$

(see (2.95)) and that

$$\dim L^{\nu} = 1 \quad \text{for all } \nu \in \Delta \tag{4.58}$$

(see b)). Thus it remains to show that $\langle L^{\lambda}_{\alpha} , L^{\mu}_{\beta} \rangle \neq \{0\}$. It is well-known that this is true if α and β are not both equal to $\bar{1}$. Suppose now that $\alpha = \beta = \bar{1}$. Let ϕ be a non-degenerate even supersymmetric invariant bilinear form on L (see d)). We choose non-zero elements

$$X_{\lambda} \in L^{\lambda}_{\bar{1}} \quad , \quad Y_{\mu} \in L^{\mu}_{\bar{1}} \quad , \quad E_{-\lambda-\mu} \in L^{-\lambda-\mu}_{\bar{0}} \; . \tag{4.59}$$

Let $H_{\lambda+\mu}$ be the element of h corresponding to the root $\lambda + \mu$ by means of the equation (3.34). Using the equation (3.37) as well as the invariance of ϕ it is not difficult to see that

$$\langle E_{-\lambda-\mu} , \langle X_{\lambda} , Y_{\mu} \rangle\rangle = \phi (X_{\lambda} , \langle E_{-\lambda-\mu} , Y_{\mu} \rangle) H_{\lambda+\mu} \; . \tag{4.60}$$

Since $-\lambda - \mu \in \Delta_{\bar{0}}$, $\mu \in \Delta_{\bar{1}}$, $-\lambda \in \Delta_{\bar{1}}$, we already know that $\langle E_{-\lambda-\mu} , Y_{\mu} \rangle$ is a non-zero element of $L_{\bar{1}}^{-\lambda}$. But the restriction of ϕ to $L_{\bar{1}}^{\lambda} \times L_{\bar{1}}^{-\lambda}$ is non-degenerate (see §3, n°2). Since both $L_{\bar{1}}^{\lambda}$ and $L_{\bar{1}}^{-\lambda}$ are one-dimensional we conclude that

$$\phi(X_{\lambda}, \langle E_{-\lambda-\mu}, Y_{\mu} \rangle) \neq 0 . \tag{4.61}$$

On the other hand $\lambda + \mu$ is an even root and hence is different from zero. But now equation (4.60) implies that

$$\langle X_{\lambda}, Y_{\mu} \rangle \neq 0 , \tag{4.62}$$

as required.

§5 CLASSIFICATION OF THE CLASSICAL SIMPLE LIE SUPERALGEBRAS

We remind the reader that all Lie superalgebras are assumed to be *finite-dimensional*. Throughout this paragraph we shall suppose that the field K is *algebraically closed*.

In the preceding paragraph we have described several families of classical simple Lie superalgebras. We shall now show that (up to isomorphism) no other classical simple Lie superalgebras do exist. As a by-product of our proof we shall also obtain a classification of a special type of transitive irreducible Z-graded Lie superalgebras (see proposition 2 below). Finally, we shall discuss the extension of some classical simple Z-graded Lie superalgebras.

Theorem 1

We suppose that the field K is algebraically closed.

A classical simple Lie superalgebra is either a simple Lie algebra or else isomorphic to one of the following classical simple Lie superalgebras :

spl(n,m) with $n, m \geq 1$; $n \neq m$

spl(n,n) / $K \cdot I_{2n}$ with $n \geq 2$

osp(n,2r) with $n, r \geq 1$

b(n) with $n \geq 3$

d(n) / $K \cdot I_{2n}$ with $n \geq 3$

$\Gamma(\sigma_1, \sigma_2, \sigma_3)$ with $\sigma_i \in K$, $\sigma_i \neq 0$, $\sigma_1 + \sigma_2 + \sigma_3 = 0$

Γ_2, Γ_3.

Remark 1)

Between the Lie superalgebras listed in theorem 1 there exist the following isomorphisms :

$$\mathrm{spl}(n,m) \simeq \mathrm{spl}(m,n) \quad \text{for all } n, m \tag{5.1}$$

$$\mathrm{spl}(2,1) \simeq \mathrm{osp}(2,2) \tag{5.2}$$

$$osp(4,2) \simeq \Gamma(-2,1,1) \quad . \tag{5.3}$$

Furthermore, we remind the reader of the isomorphisms between the various algebras $\Gamma(\sigma_1,\sigma_2,\sigma_3)$ (see §4, n°5).

Theorem 1 and the results of §4 imply :

Corollary

We suppose that the field K is algebraically closed.

A simple Lie superalgebra whose Killing form is non-degenerate is either a simple Lie algebra or else isomorphic to one of the following Lie superalgebras :

$spl(n,m)$ with $n, m \geqslant 1$; $n \neq m$

$osp(n,2r)$ with $n, r \geqslant 1$; $n \neq 2r+2$

Γ_2 , Γ_3 .

The following useful result is a by-product of the proof of theorem 1 :

Proposition 1

We suppose that the field K is algebraically closed.

Let $L_{\bar{0}}$ be a finite-dimensional Lie algebra and let ρ be a representation of $L_{\bar{0}}$ in a finite-dimensional vector space $L_{\bar{1}}$. Assume that $L_{\bar{0}}$ is not isomorphic to $sl(2) \times sl(2) \times sl(2)$. Then there exists up to a factor at most one bilinear mapping $P : L_{\bar{1}} \times L_{\bar{1}} \longrightarrow L_{\bar{0}}$ such that the Z_2 - graded vector space $L = L_{\bar{0}} \oplus L_{\bar{1}}$, equipped with the multiplication given by $L_{\bar{0}}$, ρ and P (see the equations (1.16) of chapter I), becomes a classical simple Lie superalgebra.

The proof of theorem 1 is rather lengthy, hence we shall subdivide it into several pieces [7].

1. A trivial preliminary remark

In this section we shall draw the reader's attention to some trivial process by which we can construct a "new" Lie superalgebra out of a given one. It turns out that both algebras are isomorphic. Of course, we shall classify the simple Lie superalgebras up to isomorphism; hence we have to be aware of this process to avoid a "double counting" of some algebras. For later reference, the result is formulated as a lemma.

Lemma 1

Let L be a Lie superalgebra (whose multiplication is denoted by $\langle\ ,\ \rangle$), let c be a non-zero element of K and let τ be an automorphism of the Lie algebra $L_{\bar{0}}$. Define a new superalgebra L', whose underlying Z_2-graded vector space is equal to that of L but whose multiplication $\langle\ ,\ \rangle'$ is given by

$$\langle Q_1, Q_2\rangle' = \langle Q_1, Q_2\rangle \tag{5.4,a}$$

$$\langle Q, X\rangle' = \langle \tau^{-1}(Q), X\rangle \tag{5.4,b}$$

$$\langle X, Q\rangle' = \langle X, \tau^{-1}(Q)\rangle \tag{5.4,c}$$

$$\langle X_1, X_2\rangle' = \frac{1}{c^2}\tau(\langle X_1, X_2\rangle) \tag{5.4,d}$$

for all $Q, Q_1, Q_2 \in L_{\bar{0}}$ and $X, X_1, X_2 \in L_{\bar{1}}$.

Then L' is a Lie superalgebra and the linear mapping

$$g : L \longrightarrow L' \tag{5.5,a}$$

defined by

$$g(Q) = \tau(Q) \quad \text{if } Q \in L_{\bar{0}} \tag{5.5,b}$$

$$g(X) = cX \quad \text{if } X \in L_1 \tag{5.5,c}$$

is an isomorphism of Lie superalgebras.

One should note that the representations of $L_{\bar{0}}$ in the odd subspaces of L and L' are not necessarily equivalent. Moreover, the lemma shows that a "rescaling" of the product mapping $L_{\bar{1}} \times L_{\bar{1}} \longrightarrow L_{\bar{0}}$ leads to isomorphic Lie superalgebras.

In the following, L *will denote a classical simple Lie superalgebra with a non-vanishing odd subspace.* We shall distinguish several cases depending on whether the (reductive) Lie algebra $L_{\bar{0}}$ is simple or not and on whether the representation ad' of $L_{\bar{0}}$ in $L_{\bar{1}}$ is irreducible or not.

2. $L_{\bar{0}}$ is not simple, ad' is irreducible

Since ad' is irreducible the center of $L_{\bar{0}}$ must be trivial (see the corollary to theorem 1 in §2). Hence $L_{\bar{0}}$ may be written as a direct product of two non-zero semi-simple Lie algebras $L_{\bar{0}}^1$ and $L_{\bar{0}}^2$,

$$L_{\bar{0}} = L_{\bar{0}}^1 \times L_{\bar{0}}^2 \ . \tag{5.6}$$

By assumption, ad' is irreducible; moreover, ad' is faithful (see §2, lemma 2). Hence there exists, for $i = 1, 2$, an irreducible faithful $L_{\bar{0}}^i$-module V_i such that the $L_{\bar{0}}$-module $L_{\bar{1}}$ is isomorphic to the $L_{\bar{0}}^1 \times L_{\bar{0}}^2$-module $V_1 \otimes V_2$. In the following, we shall identify $L_{\bar{1}}$ with $V_1 \otimes V_2$.
The simplicity of L implies that

$$\langle L_{\bar{1}}, L_{\bar{1}} \rangle = L_{\bar{0}} \tag{5.7}$$

(see §2, lemma 2). A short discussion of the product map $L_{\bar{1}} \times L_{\bar{1}} \longrightarrow L_{\bar{0}}$ then shows that there exist, for $i = 1, 2$, a non-degenerate $L_{\bar{0}}^i$-invariant bilinear form ψ_i on V_i and a non-zero $L_{\bar{0}}^i$-invariant bilinear mapping

$$P_i : V_i \times V_i \longrightarrow L_{\bar{0}}^i \tag{5.8}$$

such that

$$\langle u_1 \otimes u_2 , v_1 \otimes v_2 \rangle = \psi_2(u_2,v_2)\, P_1(u_1,v_1) + \psi_1(u_1,v_1)\, P_2(u_2,v_2) \tag{5.9}$$
$$\text{for all } u_i , v_i \in V_i \ ; \ i = 1, 2 \ .$$

It is well-known that ψ_i is determined up to a non-zero factor once the self-contragredient irreducible $L_{\bar{0}}^i$-module V_i is given; in particular, ψ_i must be either symmetric or skew-symmetric. Since the product mapping $L_{\bar{1}} \times L_{\bar{1}} \longrightarrow L_{\bar{0}}$ is symmetric it follows that the mappings ψ_2, P_1 must be either both symmetric or both skew-symmetric; of course, a similar statement holds true for ψ_1, P_2.

We are now going to exploit the Jacobi identity for three odd elements :

$$\langle\langle u_1 \otimes u_2 , v_1 \otimes v_2 \rangle , w_1 \otimes w_2 \rangle + \text{cyclic} = 0 \tag{5.10}$$
$$\text{for all } u_i , v_i , w_i \in V_i ; \; i = 1 , 2 .$$

For every $Q_i \in L^i_{\bar{0}}$ let $\tilde{Q}_i$ denote the corresponding homothety of the $L^i_{\bar{0}}$-module V_i . We have to distinguish two cases.

A. First we shall assume that

$$\dim V_i \geqslant 3 \quad \text{for } i = 1 , 2 . \tag{5.11}$$

Inserting the expression (5.9) into (5.10) we first see that there exist some constants $\omega_i , \sigma_i , \tau_i \in K$ such that

$$\tilde{P}_i(u_i,v_i)\, w_i = \omega_i\, \psi_i(u_i,v_i)\, w_i + \sigma_i\, \psi_i(v_i,w_i)\, u_i + \tau_i\, \psi_i(w_i,u_i)\, v_i \tag{5.12}$$
$$\text{for all } u_i , v_i , w_i \in V_i ; \; i = 1 , 2 ;$$

then we deduce that (5.10) is fulfilled if and only if

$$\omega_1 + \omega_2 = \sigma_1 + \tau_2 = \sigma_2 + \tau_1 = 0 . \tag{5.13}$$

On the other hand the bilinear form ψ_i is $L^i_{\bar{0}}$-invariant; in particular, we must have

$$\psi_i(\tilde{P}_i(u_i,v_i)\, w_i , \bar{w}_i) + \psi_i(w_i , \tilde{P}_i(u_i,v_i)\bar{w}_i) = 0 \tag{5.14}$$
$$\text{for all } u_i , v_i , w_i , \bar{w}_i \in V_i ; \; i = 1 , 2 .$$

Using the equation (5.12) this condition is satisfied if and only if

$$\omega_i = \sigma_i + \tau_i = 0 \quad \text{for } i = 1 , 2 . \tag{5.15}$$

It follows that

$$\tilde{P}_i(u_i,v_i)\, w_i = \sigma \{ \psi_i(v_i,w_i)\, u_i - \psi_i(w_i,u_i)\, v_i \} \tag{5.16}$$
$$\text{for all } u_i , v_i , w_i \in V_i ; \; i = 1 , 2$$

with some non-zero element $\sigma \in K$.

Obviously, P_i is symmetric (resp. skew-symmetric) if and only if ψ_i is skew-symmetric (resp. symmetric). Thus one of the bilinear forms ψ_i is symmetric and the other is skew-symmetric.

Without loss of generality we may assume that ψ_1 is symmetric and that ψ_2 is skew-symmetric. Then it is well-known that the linear mappings $\tilde{P}_i(u_i, v_i)$; u_i , $v_i \in V_i$; $i = 1$ (resp. $i = 2$), generate a subspace of $gl(V_i)$ which is equal to the orthogonal Lie algebra $o(\psi_1)$ (resp. to the symplectic Lie algebra $sp(\psi_2)$). But we know that ψ_i is $L^i_{\bar{0}}$ - invariant and that the representation of $L^i_{\bar{0}}$ in V_i is faithful. Therefore, the representation of $L^1_{\bar{0}}$ in V_1 (resp. of $L^2_{\bar{0}}$ in V_2) is an isomorphism of the Lie algebra $L^1_{\bar{0}}$ (resp. of $L^2_{\bar{0}}$) onto the orthogonal Lie algebra $o(\psi_1)$ (resp. onto the symplectic Lie algebra $sp(\psi_2)$).

The equations (5.9) and (5.16) show that the product map $L_{\bar{1}} \times L_{\bar{1}} \longrightarrow L_{\bar{0}}$ is fixed up to a factor once the Lie algebra $L_{\bar{0}} = L^1_{\bar{0}} \times L^2_{\bar{0}}$ and the $L^i_{\bar{0}}$-modules V_i are given. In view of lemma 1 this implies that the Lie superalgebra L must be isomorphic to an orthosymplectic algebra $osp(n,2r)$ with $n \geqslant 3$, $r \geqslant 2$.

B. Let us now consider the case where the condition (5.11) is not fulfilled. Since the $L^i_{\bar{0}}$ - module V_i is faithful we conclude that (at least) one of the spaces V_i is 2 - dimensional and that the corresponding Lie algebra $L^i_{\bar{0}}$ is isomorphic to $sl(2)$. Without loss of generality we may assume that

$$\dim V_2 = 2 \quad , \quad L^2_{\bar{0}} = sl(V_2) \tag{5.17}$$

and that V_2 is the elementary $L^2_{\bar{0}}$ - module. It is well-known that there exist a non-degenerate skew-symmetric $L^2_{\bar{0}}$ - invariant bilinear form ψ_2 on V_2 and a non-zero $L^2_{\bar{0}}$ - invariant bilinear mapping

$$P_2 : V_2 \times V_2 \longrightarrow L^2_{\bar{0}} . \tag{5.18}$$

Both ψ_2 and P_2 are fixed up to a factor, in particular, we have

$$\tilde{P}_2(u_2, v_2)\, w_2 = \sigma \{ \psi_2(v_2, w_2)\, u_2 - \psi_2(w_2, u_2)\, v_2 \} \tag{5.19}$$
$$\text{for all } u_2 , v_2 , w_2 \in V_2 ,$$

where σ is some non-zero element of K.

Therefore, the mappings ψ_2 and P_2 in equation (5.9) are already known. We conclude that the bilinear form ψ_1 is symmetric, i.e. that the $L_{\bar{0}}^1$-module V_1 is orthogonal; moreover, the mapping P_1 is skew-symmetric.

Now it is not difficult to show that the Jacobi identity (5.10) is fulfilled if and only if the $L_{\bar{0}}^1$-invariant trilinear mapping

$$\hat{P}_1 : V_1 \times V_1 \times V_1 \longrightarrow V_1 \tag{5.20,a}$$

defined by

$$\hat{P}_1(u_1,v_1,w_1) = \tilde{P}_1(u_1,v_1)w_1 - \sigma\{\psi_1(v_1,w_1)u_1 - \psi_1(w_1,u_1)v_1\} \quad \text{for all } u_1, v_1, w_1 \in V_1 \tag{5.20,b}$$

is totally skew-symmetric.

In the case $\hat{P}_1 = 0$ we are back at equation (5.16) and we can conclude that L must be isomorphic to an orthosymplectic algebra osp(n,2) with $n \geq 3$. But it turns out that $\hat{P}_1$ does not necessarily vanish. Nevertheless, it is obvious that $\hat{P}_1 = 0$ if $n \leq 3$. Under the present assumptions the case $n = 2$ is impossible; for $n = 3$ we obtain the algebra osp(3,2).

Let us next consider the case $n = 4$. Then $L_{\bar{0}}^1$ is a semi-simple Lie algebra which has an irreducible faithful orthogonal 4-dimensional representation. It follows that $L_{\bar{0}}^1 \simeq sl(2) \times sl(2) \simeq o(4)$ and that V_1 is the elementary o(4)-module. This is one of the (exceptional) cases in which $\hat{P}_1$ is not necessarily equal to zero. However, this case has been treated quite generally in chapter I, §1, example 5) : We know that L is isomorphic to one of the exceptional Lie superalgebras $\Gamma(\sigma_1,\sigma_2,\sigma_3)$ (which include osp(4,2)).

Thus we now may assume that

$$\dim V_1 \geq 5 \tag{5.21}$$

and that the Lie algebra $L_{\bar{0}}^1$ is *simple* (the case where $L_{\bar{0}}^1$ is not simple leads at once back to case A).

Under these assumptions the Killing form of L is non-degenerate. In fact, it is easy to see that the restriction to $L_{\bar{0}}^2$ of the Killing form of L is equal to $1-\frac{1}{4}\dim V_1$ times the Killing form of $L_{\bar{0}}^2$; thus the Killing form of L is non-zero, hence it is non-degenerate (see §2, proposition 2).

Therefore, we now may apply the results of §3, n°2 ; we shall use the notation introduced in that section. Let μ be one of the two weights of the $L_{\bar{0}}^2$ - module V_2 . Then the weights of ad' are exactly the linear forms of the type $\alpha = (\tilde{\alpha}, \pm\mu)$ where $\tilde{\alpha}$ is a weight of the $L_{\bar{0}}^1$ - module V_1 . Let us normalize the invariant bilinear form ϕ on L in such a way that

$$(\mu|\mu) = -1 \ . \tag{5.22}$$

If $\tilde{\alpha} \neq 0$, then 2α is not a root of $L_{\bar{0}}$; hence the lemma 2 of §3 implies that $(\alpha|\alpha) = 0$, i.e. that

$$(\tilde{\alpha}|\tilde{\alpha}) = 1 \ . \tag{5.23}$$

This equation shows that the restriction of ϕ to $L_{\bar{0}}^1$ is a positive rational multiple of the Killing form of $L_{\bar{0}}^1$.

Suppose now that $\tilde{\alpha}$ and $\tilde{\beta}$ are two weights of the $L_{\bar{0}}^1$ - module V_1 . Then

$$\alpha = (\tilde{\alpha}, \mu) \quad , \quad \beta = (\tilde{\beta}, \mu) \tag{5.24}$$

are two weights of ad' and we have

$$(\alpha|\beta) = (\tilde{\alpha}|\tilde{\beta}) + (\mu|\mu) = (\tilde{\alpha}|\tilde{\beta}) - 1 \ . \tag{5.25}$$

Because of the equation (5.23) this expression will vanish if and only if $\tilde{\alpha} = \tilde{\beta} \neq 0$.

Now let $\tilde{\alpha} \neq \pm\tilde{\beta}$; then $\alpha + \beta$ is not a root of $L_{\bar{0}}$, hence $\alpha - \beta = (\tilde{\alpha} - \tilde{\beta}, 0)$ must be a root of $L_{\bar{0}}$ (see §3, lemma 3). Thus we have shown :

<u>Lemma</u> 2

If $\tilde{\alpha}$ and $\tilde{\beta}$ are two weights of the $L_{\bar{0}}^1$ - module V_1 such that $\tilde{\alpha} \neq \pm\tilde{\beta}$, then $\tilde{\alpha} - \tilde{\beta}$ is a root of $L_{\bar{0}}^1$.

Now recall that the $L_{\bar{0}}^1$ - module V_1 is orthogonal and that the inequality (5.21) is assumed to hold. Therefore, according to lemma 2 (see also remark 2)) of the appendix we are left with the following possibilities :

a) $L_{\bar{0}}^{1} \simeq o(n)$ with some $n \geq 5$ and V_1 is the elementary $o(n)$ - module.

a') $L_{\bar{0}}^{1} \simeq o(8)$ and V_1 carries one of the two 8 - dimensional half-spin representations of $o(8)$.

b) $L_{\bar{0}}^{1} \simeq G_2$ and V_1 is the 7 - dimensional fundamental G_2 - module.

c) $L_{\bar{0}}^{1} \simeq o(7)$ and V_1 carries the 8 - dimensional spin representation of $o(7)$.

Note that the elementary representation $\rho(\lambda_1)$ and the two half-spin representations $\rho(\lambda_3)$ and $\rho(\lambda_4)$ of $o(8) \simeq D_4$ are connected by suitable automorphisms of the Lie algebra D_4 . Therefore, in view of lemma 1 , the cases a') may be dropped in favour of case a) .

In all these cases there exists one and, up to a constant factor, only one non-zero $L_{\bar{0}}^{1}$ - invariant bilinear mapping $P_1 : V_1 \times V_1 \longrightarrow L_{\bar{0}}^{1}$. It is obvious from equation (5.20,b) that for at most one choice of the free factor the mapping $\hat{P}_1$ will be totally skew-symmetric. Once again, therefore, the product mapping $L_{\bar{1}} \times L_{\bar{1}} \longrightarrow L_{\bar{0}}$ is fixed up to a factor if the Lie algebra $L_{\bar{0}}$ and the $L_{\bar{0}}^{i}$ - modules V_i are given. Consequently (recall lemma 1) the Lie superalgebra L must be isomorphic to one of the algebras osp(n,2) ; $n \geq 5$, or Γ_2 , Γ_3 .

3. $L_{\bar{0}}$ <u>is not simple,</u> ad' <u>is not irreducible</u>

This case can be treated by a procedure which is completely analogous to the one used in the previous section. However, we shall modify the argument so as to prove the following more general proposition.

<u>Proposition</u> 2

We suppose that the field K is algebraically closed.

Let $G = \bigoplus_{i \geq -1} G_i$ be a transitive irreducible consistently Z - graded Lie superalgebra satisfying the following conditions :

a) The representations of G_0 in G_{-1} and G_1 are contragredient to each other.

b) The subspace $G_{-1} \oplus G_1$ generates the algebra G .

Then the Z - graded Lie superalgebra G is isomorphic to one of the following Z - graded Lie superalgebras :

spl(n,m) with $n > m \geqslant 1$

spl(n,n) / $K \cdot I_{2n}$ with $n \geqslant 2$

osp(2,2r) with $r \geqslant 1$.

(Recall that spl(2,1) and osp(2,2) are isomorphic.)

Let us first show that this proposition settles that part of theorem 1 which is mentioned in the headline to the present section. Set $L_{\bar{0}} = G_0$. Since the representation ad' is completely reducible but (by assumption) not irreducible we infer from proposition 3 of §2 that $L_{\bar{1}}$ decomposes into the direct sum of two G_0 - irreducible subspaces G_{-1} and G_1 such that

$$\langle G_{-1}, G_1 \rangle = G_0 \ . \tag{5.26}$$

Furthermore, we know that $(G_i)_{-1 \leqslant i \leqslant 1}$ is a transitive Z - gradation of the Lie superalgebra L which is consistent with the Z_2 - gradation (see proposition 3, remark 4) and lemma 4 of §2).

Thus it remains to show that the G_0 - modules G_{-1} and G_1 are contragredient to each other. But this follows from equation (5.26) and our assumption that the (reductive) Lie algebra $G_0 = L_{\bar{0}}$ is not simple. In fact, one may either argue directly (using the corollary to theorem 1 of §2 or the proposition 4 of §3) or else one may invoke the proposition 4 of §1 .

Proof of proposition 2

By assumption the Z - graded Lie superalgebra G is transitive and irreducible, moreover, the representations of G_0 in G_{-1} and G_1 are contragredient to each other. It follows that both of these representations are irreducible and faithful (see §1, lemma 4) and that the Z - graded Lie superalgebra G is *bitransitive* (see §1, lemma 1).

On the other hand, let us recall (see §1, proposition 3) that the Lie algebra G_0 is reductive (but non-abelian) and that the center G_0^o of G_0 is at most one-dimensional. Moreover, if $\dim G_0^o = 1$, then there exists a unique element $C \in G_0^0$ such that for all $j \geqslant -1$

$$\langle C , X \rangle = j X \quad \text{if } X \in G_j . \tag{5.27}$$

Since the subspace $G_{-1} \oplus G_1$ generates the algebra G we conclude that

$$\langle G_{-1} , G_1 \rangle = G_0 . \tag{5.28}$$

After these preliminaries we shall distinguish two cases.

A. $\langle G_0 , G_0 \rangle$ <u>is not simple</u>

In this case G_0 may be written as a direct product

$$G_0 = G_0^0 \times G_0^1 \times G_0^2 \tag{5.29}$$

where G_0^1 and G_0^2 are two non-zero semi-simple Lie algebras. The $G_0^1 \times G_0^2$-modules G_{-1} and G_1 are irreducible, faithful, and contragredient to each other. Hence there exist, for $i = 1 , 2$, two irreducible faithful G_0^i-modules U_i and V_i such that the $G_0^1 \times G_0^2$-modules G_{-1} and $U_1 \otimes U_2$ (resp. G_1 and $V_1 \otimes V_2$) are isomorphic. Furthermore, the two G_0^i-modules U_i and V_i are contragredient to each other, i.e. there exists, for $i = 1 , 2$, a non-degenerate G_0^i-invariant bilinear form ψ_i on $U_i \times V_i$. It is well-known that ψ_i is fixed up to a factor once the two irreducible contragredient G_0^i-modules U_i and V_i are given.

In the following we shall identify G_{-1} with $U_1 \otimes U_2$ and G_1 with $V_1 \otimes V_2$. Then equation (5.28) implies that there exists, for $i = 1 , 2$, a non-zero G_0^i-invariant bilinear mapping

$$P_i : U_i \times V_i \longrightarrow G_0^i \tag{5.30}$$

such that

$$\begin{aligned} &\langle u_1 \otimes u_2 , v_1 \otimes v_2 \rangle \\ &= \psi_2(u_2,v_2)\, P_1(u_1,v_1) + \psi_1(u_1,v_1)\, P_2(u_2,v_2) + \psi_1(u_1,v_1)\, \psi_2(u_2,v_2)\, F \\ &\qquad \text{for all } u_i \in U_i \text{ and } v_i \in V_i \; ; \; i = 1 , 2 . \end{aligned} \tag{5.31}$$

Here F is a suitable element of G_0^0 such that $F \neq 0$ if $G_0^0 \neq \{0\}$. It is convenient to define a constant $\eta \in K$ by

$$\eta = 0 \quad \text{if } G_0^0 = \{0\} \tag{5.32,a}$$

$$F = \eta C \quad \text{if } G_0^0 \neq \{0\} \tag{5.32,b}$$

(see equation (5.27)). Note that $\eta = 0$ if and only if $G_0^0 = \{0\}$.

For every $Q_i \in G_0^i$ let $\tilde{Q}_i$ (resp. $\hat{Q}_i$) denote the corresponding homothety of the G_0^i - module U_i (resp. V_i). The Jacobi identity for two elements from G_{-1} and one element from G_1 is equivalent to

$$\langle\langle u_1 \otimes u_2 , v_1 \otimes v_2 \rangle , \bar{u}_1 \otimes \bar{u}_2 \rangle + \langle\langle \bar{u}_1 \otimes \bar{u}_2 , v_1 \otimes v_2 \rangle , u_1 \otimes u_2 \rangle = 0 \tag{5.33}$$
$$\text{for all } u_i , \bar{u}_i \in U_i \text{ and } v_i \in V_i ; \ i = 1 , 2 .$$

Inserting the expression (5.31) into (5.33) we see first that there exist some constants $\sigma_i , \tau_i \in K$ such that

$$\tilde{P}_i(u_i,v_i)\,\bar{u}_i = \sigma_i\, \psi_i(u_i,v_i)\,\bar{u}_i + \tau_i\, \psi_i(\bar{u}_i,v_i)\,u_i \tag{5.34}$$
$$\text{for all } u_i , \bar{u}_i \in U_i \text{ and } v_i \in V_i ; \ i = 1 , 2 ;$$

then we deduce that (5.33) is fulfilled if and only if

$$\sigma_1 + \sigma_2 - \eta = \tau_1 + \tau_2 = 0 . \tag{5.35}$$

On the other hand the bilinear form ψ_i is G_0^i - invariant; in particular, we must have

$$\psi_i(\tilde{P}_i(u_i,v_i)\,\bar{u}_i , \bar{v}_i) + \psi_i(\bar{u}_i , \hat{P}_i(u_i,v_i)\,\bar{v}_i) = 0 \tag{5.36}$$
$$\text{for all } u_i , \bar{u}_i \in U_i \text{ and } v_i , \bar{v}_i \in V_i ; \ i = 1 , 2 .$$

This implies that

$$\hat{P}_i(u_i,v_i)\,\bar{v}_i = -\sigma_i\, \psi_i(u_i,v_i)\,\bar{v}_i - \tau_i\, \psi_i(u_i,\bar{v}_i)\,v_i \tag{5.37}$$
$$\text{for all } u_i \in U_i \text{ and } v_i , \bar{v}_i \in V_i ; \ i = 1 , 2 .$$

Now it is easy to check that

$$\langle u_1 \otimes u_2 , \langle v_1 \otimes v_2 , \bar{v}_1 \otimes \bar{v}_2 \rangle\rangle \tag{5.38}$$
$$= \langle\langle u_1 \otimes u_2 , v_1 \otimes v_2 \rangle , \bar{v}_1 \otimes \bar{v}_2 \rangle + \langle\langle u_1 \otimes u_2 , \bar{v}_1 \otimes \bar{v}_2 \rangle , v_1 \otimes v_2 \rangle = 0$$
$$\text{for all } u_i \in U_i \text{ and } v_i , \bar{v}_i \in V_i ; \ i = 1 , 2 .$$

This means that

$$\langle G_{-1}, \langle G_1 , G_1 \rangle\rangle = \{0\} . \tag{5.39}$$

But the Z - graded Lie superalgebra G is transitive, hence the equation (5.39) implies that

$$\langle G_1 , G_1 \rangle = \{0\} . \tag{5.40}$$

Since the subspace $G_{-1} \oplus G_1$ generates the algebra G we conclude that

$$G = G_{-1} \oplus G_0 \oplus G_1 . \tag{5.41}$$

Let us now recall that for any finite-dimensional module over a semi-simple Lie algebra the homotheties are traceless. In particular, we have

$$\operatorname{Tr} \tilde{P}_i(u_i , v_i) = - \operatorname{Tr} \hat{P}_i(u_i , v_i) = 0 \tag{5.42}$$

for all $u_i \in U_i$ and $v_i \in V_i$; $i = 1 , 2$.

If we set

$$n_i = \dim U_i = \dim V_i \quad ; \quad i = 1 , 2 \tag{5.43}$$

this condition is equivalent to

$$n_i \sigma_i + \tau_i = 0 \quad ; \quad i = 1 , 2 . \tag{5.44}$$

Considering the dimensions n_1 , n_2 as fixed, we can rephrase the equations (5.35) and (5.44) by demanding that there should exist a non-zero element $\tau \in K$ such that

$$\begin{aligned} \sigma_1 &= \frac{\tau}{n_1} , & \tau_1 &= - \tau \\ \sigma_2 &= -\frac{\tau}{n_2} , & \tau_2 &= \tau \\ & & \eta &= \frac{n_2 - n_1}{n_2 n_1} \tau . \end{aligned} \tag{5.45}$$

Note that $\eta = 0$ if and only if $n_1 = n_2$.

It is now easy to see that, for $i = 1 , 2$, the mappings $\tilde{P}_i(u_i , v_i)$ (resp. $\hat{P}_i(u_i , v_i)$); $u_i \in U_i$, $v_i \in V_i$, generate the subspace $sl(U_i)$ of $gl(U_i)$

(resp. $sl(V_i)$ of $gl(V_i)$). On the other hand, the Lie algebra G_0^i is semi-simple and the G_0^i-modules U_i and V_i are faithful. All this implies that the representation of G_0^i in U_i (resp. in V_i) is an isomorphism of the Lie algebra G_0^i onto $sl(U_i)$ (resp. onto $sl(V_i)$).

The equations (5.31), (5.34), (5.37) and (5.45) show that the product mapping $G_{-1} \times G_1 \longrightarrow G_0$ is fixed up to a factor once the Lie algebras G_0^i and the contragredient G_0^i-modules U_i and V_i are given. In view of lemma 1 this implies that the Z-graded Lie superalgebra G is isomorphic to $spl(n_1,n_2)$ if $n_1 \neq n_2$ but isomorphic to $spl(n_1,n_1)/K\cdot I_{2n_1}$ if $n_1 = n_2$. Note that according to our assumptions $n_1, n_2 \geqslant 2$.

B. $\langle G_0, G_0 \rangle$ is simple

By assumption we know that

$$G_0 = G_0^0 \times G_0^1 \tag{5.46}$$

where $G_0^1 = \langle G_0, G_0 \rangle$ is a simple Lie algebra. Let us choose a Cartan subalgebra h of G_0 as well as a fundamental system of simple roots of G_0 with respect to h. Furthermore, let ϕ be a non-degenerate symmetric invariant bilinear form on G_0. We shall use the notions introduced in the appendix. In particular, we have $h = G_0^0 \times h^1$ where h^1 is a Cartan subalgebra of G_0^1. Moreover, we define, for every root α of G_0, the element $H_\alpha \in h$ by

$$\alpha(H') = \phi(H_\alpha, H') \quad \text{for all } H' \in h \tag{5.47}$$

and we choose the root vectors $E_{\pm\alpha}$ associated with the roots $\pm\alpha$ such as to satisfy

$$\langle E_{-\alpha}, E_\alpha \rangle = H_\alpha \quad \text{for all roots } \alpha . \tag{5.48}$$

Let λ (resp. μ) be the highest (resp. lowest) weight of the G_0-module G_{-1} (resp. G_1) and let $X_\lambda \in G_{-1}$ (resp. $Y_\mu \in G_1$) be a weight vector associated with it. Set

$$\langle X_\lambda, Y_\mu \rangle = H \; ; \tag{5.49}$$

we know (see §1, proposition 4) that

$$\mu = -\lambda \tag{5.50}$$

and that

$$H \in h \quad , \quad H \notin G_0^0 \, . \tag{5.51}$$

Lemma 3

We have

$$\lambda(H) = 0 \, . \tag{5.52}$$

Proof

Obviously, $\langle X_\lambda , X_\lambda \rangle = 0$. On the other hand it is easy to check that

$$\langle Y_{-\lambda} , \langle X_\lambda , X_\lambda \rangle\rangle = 2\lambda(H) X_\lambda \, , \tag{5.53}$$

hence equation (5.52) is valid.

Lemma 4

If α is any positive root of G_0 then

$$\alpha(H)(\alpha|\lambda)(\alpha|2\lambda-\alpha) = 0 \, . \tag{5.54}$$

Proof

Consider the element

$$A = \langle\langle E_{-\alpha} , X_\lambda\rangle , \langle E_{-\alpha} , X_\lambda\rangle\rangle \, . \tag{5.55}$$

Obviously, we have $A = 0$. On the other hand it is easy to see that

$$\langle E_\alpha , \langle E_\alpha , \langle Y_{-\lambda} , A\rangle\rangle\rangle = 2\alpha(H)(\alpha|2\lambda-\alpha)(\alpha|\lambda) X_\lambda \, . \tag{5.56}$$

This implies equation (5.54).

Lemma 5

Let α and β be two different positive roots of G_0 . Suppose that $\alpha - \beta$ is not a root of G_0 . Then

$$\begin{aligned} (\alpha+\beta)(H)(\alpha|\lambda)(\beta|\lambda) &= (\alpha|\beta)\,\alpha(H)(\beta|\lambda) \\ &= (\alpha|\beta)\,\beta(H)(\alpha|\lambda) \, . \end{aligned} \tag{5.57}$$

Proof

Consider the element

$$B = \langle X_\lambda, \langle E_{-\alpha}, \langle E_{-\beta}, X_\lambda \rangle\rangle\rangle . \tag{5.58}$$

We know that $B = 0$. On the other hand it is not difficult to check that

$$\langle E_\beta, \langle E_\alpha, \langle Y_{-\lambda}, B \rangle\rangle\rangle = -\{(\alpha+\beta)(H)(\alpha|\lambda) - (\alpha|\beta)\alpha(H)\}(\beta|\lambda) X_\lambda . \tag{5.59}$$

This implies the first of the equations (5.57) ; the second is deduced from the first one by interchanging α and β.

Corollary

Let α and β be two positive roots of G_0 such that $\alpha+\beta$ is a root of G_0 but $\alpha-\beta$ is not a root. Then

$$\alpha(H)(\beta|\lambda) = \beta(H)(\alpha|\lambda) . \tag{5.60}$$

Proof

Our assumptions imply that $(\alpha|\beta) \neq 0$; thus the assertion follows from the preceding lemma.

Recall that

$$h = G_0^0 \times h^1 \quad ; \quad h^* = (G_0^0)^* \times (h^1)^* . \tag{5.61}$$

Let H^1 denote the component of H in h^1 and let λ^1 denote the component of λ in $(h^1)^*$.

Lemma 6

There exists a unique non-zero element $c \in K$ such that

$$\tau(H^1) = c(\tau|\lambda^1) \quad \text{for all } \tau \in (h^1)^* . \tag{5.62}$$

Proof

Since the representation of G_0 in G_{-1} is faithful we know that $\lambda^1 \neq 0$. Hence there exists a simple root α of G_0^1 such that

$$(\alpha|\lambda^1) \neq 0 . \tag{5.63}$$

If the element c exists at all we must have

$$\alpha(H^1) = c(\alpha|\lambda^1) . \qquad (5.64)$$

Accordingly, we define the element $c \in K$ by the equation (5.64). Let γ be any simple root of G_0^1 which is different from α. Then there exists a finite sequence

$$\alpha = \beta_0 , \beta_1 , \ldots\ldots , \beta_p = \gamma \qquad (5.65)$$

of different simple roots β_q of G_0^1 whose vertices in the Dynkin diagram form a connected chain. We show by induction that

$$\beta_q(H^1) = c(\beta_q|\lambda^1) \qquad (5.66)$$

if $0 \leq q \leq p$.

By definition, (5.66) is valid if $q = 0$. Let $1 \leq r \leq p$ and suppose that (5.66) is known for $0 \leq q \leq r-1$. It is well-known that

$$\beta = \sum_{s=1}^{r} \beta_s \qquad (5.67)$$

is a positive root of G_0^1; moreover, $\alpha + \beta$ is a root of G_0^1 but $\alpha - \beta$ is not. Using the corollary to lemma 5 as well as the induction hypothesis it is easy to see that equation (5.66) holds for $q = r$.

Thus we have shown that equation (5.62) is valid if τ is a simple root of G_0^1; this implies that (5.62) holds in general.

It is now easy to see that $c \neq 0$. For suppose that $c = 0$. Then equation (5.62) implies that $H^1 = 0$ and hence that $H \in G_0^0$; but this is not the case.

Corollary

The center G_0^0 of G_0 is not equal to $\{0\}$.

Proof

Suppose the contrary. Then equation (5.62) reads

$$\tau(H) = c(\tau|\lambda) \quad \text{for all } \tau \in h^* . \qquad (5.68)$$

In view of lemma 3 this implies $c(\lambda|\lambda) = 0$, a contradiction.

Suppose now that α and β are two different positive roots of G_0^1 such that $\alpha - \beta$ is not a root. Using the equation (5.62) we deduce from equation (5.57) that

$$\{(\alpha+\beta|\lambda^1) - (\alpha|\beta)\}(\alpha|\lambda^1)(\beta|\lambda^1) = 0 . \tag{5.69}$$

On the other hand, the equations (5.54) and (5.62) combine to yield

$$(\gamma|\lambda^1)(\gamma|2\lambda^1 - \gamma) = 0 \tag{5.70}$$

for every positive root γ of G_0^1. Hence equation (5.69) may be rewritten in the form

$$(\alpha - \beta|\alpha - \beta)(\alpha|\lambda^1)(\beta|\lambda^1) = 0 . \tag{5.71}$$

By assumption, α and β are different roots and hence $(\alpha - \beta|\alpha - \beta) \neq 0$. Thus we have proved :

Lemma 7

Suppose that α and β are two different positive roots of G_0^1 such that $\alpha - \beta$ is not a root. Then $(\alpha|\lambda^1) = 0$ or $(\beta|\lambda^1) = 0$.

Corollary

λ^1 is a fundamental weight of G_0^1 which belongs to an extremal vertex of the Dynkin diagram.

Proof

Let α be a simple root of G_0^1 such that $(\alpha|\lambda^1) \neq 0$ and let β be a simple root of G_0^1 which is different from α. Then $\alpha - \beta$ is not a root of G_0^1 and consequently $(\beta|\lambda^1) = 0$. On the other hand, equation (5.70) shows that

$$2\,\frac{(\lambda^1|\alpha)}{(\alpha|\alpha)} = 1 . \tag{5.72}$$

Hence λ^1 is a fundamental weight of G_0^1.

Evidently, λ^1 and α belong to the same vertex v of the Dynkin diagram of G_0^1. Suppose now that v is not extremal. Choose two different vertices v' and v" of the Dynkin diagram which are direct neighbours of v; let β and γ be the two simple roots of G_0^1 which belong to v' and v", respectively. Then $\alpha+\beta$ and $\alpha+\gamma$ are two different positive roots of G_0^1 and $(\alpha+\beta)-(\alpha+\gamma)$ is not a root. On the other hand, $(\alpha+\beta|\lambda^1)$ and $(\alpha+\gamma|\lambda^1)$ are both equal to $(\alpha|\lambda^1)$ and hence different from 0. This is contrary to lemma 7.

Lemma 8

The condition in lemma 7 implies that

$$G_0^1 \simeq A_n \quad \text{and} \quad \lambda^1 = \lambda_1 \text{ or } \lambda^1 = \lambda_n$$

or else

$$G_0^1 \simeq C_n \quad \text{and} \quad \lambda^1 = \lambda_1 ,$$

where $n \geqslant 1$ in both cases. (See the appendix, n°3, for the enumeration of the fundamental weights λ_i.)

Proof

We know from the corollary to lemma 7 that λ^1 is a fundamental weight of G_0^1 which belongs to an extremal vertex of the Dynkin diagram.

Let (G_0^1,λ^1) be any pair consisting of a simple Lie algebra G_0^1 and a fundamental weight λ^1 as described above. Let α be the simple root of G_0^1 which corresponds to λ^1, i.e.

$$(\alpha|\lambda^1) \neq 0 \tag{5.73}$$

and let θ be the highest root of G_0^1. If the pair (G_0^1,λ^1) is not mentioned in the statement of lemma 8 then it can be excluded by applying lemma 7: One chooses $\beta = \theta$ or $\beta = \theta - \alpha$. We shall not go into the details.

We are now ready to complete the proof of proposition 2. For the Lie algebras G_0^1 and the (contragredient irreducible) G_0^1-modules $G_{\pm1}$ at hand (see lemma 8) it is well-known that there exist a non-degenerate G_0^1-invariant bilinear form ψ on $G_{-1} \times G_1$ as well as a non-zero G_0^1-invariant bilinear mapping $P: G_{-1} \times G_1 \longrightarrow G_0^1$; furthermore, both ψ and P are unique up to a factor.

According to the corollary to lemma 6 we have

$$G_0 = G_0^0 \times G_0^1 \tag{5.74}$$

where the center G_0^0 of G_0 is one-dimensional. Let C be the element of G_0^0 which has been introduced in equation (5.27). Then there exist non-zero elements $\sigma, \tau \in K$ such that

$$\langle X, Y\rangle = \sigma P(X,Y) + \tau\psi(X,Y)C \quad \text{for all } X \in G_{-1},\ Y \in G_1 . \tag{5.75}$$

For every element $Q \in G_0^1$ let $\tilde{Q}$ denote the corresponding homothety of the G_0^1-module G_{-1}. Then the Jacobi identity for two elements from G_{-1} and one element from G_1 is equivalent to the requirement that, for all $Y \in G_1$, the bilinear mapping

$$(X,\bar{X}) \longrightarrow \sigma\tilde{P}(X,Y)\bar{X} - \tau\psi(X,Y)\bar{X} \quad \text{with } X, \bar{X} \in G_{-1} \tag{5.76}$$

should be skew-symmetric.

Obviously, for every $\sigma \in K$ there will exist at most one element $\tau \in K$ such that this criterion is satisfied. Thus we have shown that the product mapping $G_{-1} \times G_1 \longrightarrow G_0$ of our Z-graded Lie superalgebra $G = \bigoplus_{i \geqslant -1} G_i$ is fixed up to a factor once the Lie algebra G_0^1 and the G_0^1-modules $G_{\pm 1}$ are chosen according to lemma 8. It now follows from the corollary to proposition 1 of §1 that the Z-graded Lie superalgebra G is isomorphic to one of the Z-graded Lie superalgebras spl(n+1,1) or osp(2,2n) with $n \geqslant 1$. (Instead of invoking the corollary mentioned above we may argue more directly: We first construct the mapping P explicitly and then check that $\langle G_{-1}, \langle G_1, G_1\rangle\rangle = \{0\}$.)

Remark 2)

If we only wanted to prove theorem 1 the discussion of the present subsection could be greatly simplified. In fact, in this case we would know that the center of $G_0 = L_{\bar{0}}$ is non-trivial and hence that the Killing form of L is non-degenerate (see §3, proposition 4). But then it is easy to see that the difference of any two different weights of the G_0-module G_{-1} is a root of G_0 (see §3, lemma 3). According to lemma 2 of the appendix this implies that lemma 8 above is valid.

4. $L_{\bar{0}}$ is simple

We shall now settle the remaining part of the proof of theorem 1. Let ϕ (resp. $\tilde{\phi}$) denote the Killing form of the Lie superalgebra L (resp. of the Lie algebra $L_{\bar{0}}$) and let ℓ denote the index of the representation ad' of $L_{\bar{0}}$ in $L_{\bar{1}}$ (see appendix, n°5). By definition, we have

$$\phi(Q,Q') = (1-\ell)\,\tilde{\phi}(Q,Q') \quad \text{for all } Q\,,Q' \in L_{\bar{0}}\,. \tag{5.77}$$

As we know (see §2, proposition 2) the Killing form of L is either non-degenerate or equal to zero. In view of equation (5.77) this means that either the Killing form of L is non-degenerate (and hence $\ell \neq 1$) or else we have $\ell = 1$ (and the Killing form of L is equal to zero). Thus we are led to distinguish the following cases.

A. The Killing form of L is non-degenerate

In this case we may apply the notation and results of §3, n°2. Let λ be any non-zero weight of the representation ad'. Since the restriction of ϕ to $L_{\bar{0}}$ is a non-zero multiple of $\tilde{\phi}$ we conclude that $(\lambda|\lambda) \neq 0$. According to lemma 2 of §3 this implies that 2λ is a root of $L_{\bar{0}}$ and that the weight space associated with λ is one-dimensional. It follows from lemma 2 of the appendix (see also equation (2.10)) that

$$L_{\bar{0}} \simeq C_n \quad ; \quad \text{ad}' \simeq \rho(\lambda_1) \tag{5.78}$$

for some $n \geqslant 1$. But then it is well-known that any two $L_{\bar{0}}$ - invariant bilinear mappings $L_{\bar{1}} \times L_{\bar{1}} \longrightarrow L_{\bar{0}}$ are proportional. Thus we have shown (see lemma 1) that L must be isomorphic to osp(1,2n).

B. $\ell = 1$ and ad' is irreducible

Since the index ℓ of the irreducible representation ad' is equal to 1 we conclude from the appendix, n°5, that ad' must be equivalent to the adjoint representation of $L_{\bar{0}}$. Moreover, we know that the product mapping $L_{\bar{1}} \times L_{\bar{1}} \longrightarrow L_{\bar{0}}$ is symmetric and $L_{\bar{0}}$ - invariant. Thus we need the solution to the following problem :

Find all simple Lie algebras g such that the adjoint representation ad_g of g is contained in the symmetric tensor product of ad_g with itself.

It turns out that our requirement is fulfilled if and only if $g \simeq A_n$ with $n \geqslant 2$; moreover, in this case the symmetric tensor product of ad_g with itself contains ad_g only once. In fact, this result may be read off from the tables contained in [31]; on the other hand, one may also use a version of Steinberg's formula for the decomposition of the tensor product of two irreducible representations [32].

In view of lemma 1 we conclude that L must be isomorphic to the (f,d) algebra $d(n+1) / K \cdot I_{2n+2}$.

C. $\ell = 1$ <u>and</u> ad' <u>is reducible</u>

In this case we know (see §2, proposition 3) that $L_{\bar{1}}$ decomposes into the direct sum of two irreducible $L_{\bar{0}}$ - submodules $L_{\bar{1}}^1$ and $L_{\bar{1}}^2$,

$$L_{\bar{1}} = L_{\bar{1}}^1 \oplus L_{\bar{1}}^2 . \qquad (5.79)$$

Moreover, the equation $\langle L_{\bar{0}}, L_{\bar{1}} \rangle = L_{\bar{1}}$ (see §2, lemma 2) implies that the $L_{\bar{0}}$ - modules $L_{\bar{1}}^i$; $i = 1 , 2$, are not trivial.

Let ℓ_i ; $i = 1 , 2$, denote the index of the representation of $L_{\bar{0}}$ in $L_{\bar{1}}^i$. Then the above remarks show that

$$\ell_1 + \ell_2 = \ell = 1 \quad \text{and} \quad \ell_1 , \ell_2 > 0 . \qquad (5.80)$$

Hence we are faced with the following problem :

Let g be a simple Lie algebra. Find all pairs of non-trivial irreducible representations of g the sum of whose indices is equal to one.

In the following we shall separately investigate all simple Lie algebras. Using the table 2 of the appendix, $n^o 5$, we shall give all "admissible pairs" of irreducible representations (in terms of the highest weights) and discuss which of these pairs lead to a simple Lie superalgebra.

Case A_n , $n \geq 1$

This case is the most complicated one. The highest weights of the admissible pairs of representations are the following :

1)	$2\lambda_1$	λ_{n-1}	$n \geq 2$
1')	$2\lambda_n$	λ_2	$n \geq 2$
2)	$2\lambda_1$	λ_2	$n \geq 2$
2')	$2\lambda_n$	λ_{n-1}	$n \geq 2$
3)	λ_3	λ_3	$n = 5$
4)	λ_1	λ_3	$n = 7$
4')	λ_7	λ_5	$n = 7$
5)	λ_1	λ_5	$n = 7$
5')	λ_7	λ_3	$n = 7$

The "primed possibilities" are connected with the non-primed ones by an automorphism of A_n . In view of lemma 1 the primed cases may, therefore, be disregarded.

1) The tensor product of $\rho(2\lambda_1)$ with $\rho(\lambda_{n-1})$ contains the adjoint representation of A_n exactly once. According to lemma 1 the corresponding simple Lie superalgebra L is isomorphic to $b(n+1)$.

2) We may assume that $n \neq 3$ since the case $n = 3$ is included in 1). Then the tensor product of $\rho(2\lambda_1)$ with $\rho(\lambda_2)$ does not contain the adjoint representation; hence this case does not lead to a simple Lie superalgebra.

3) The tensor product of $\rho(\lambda_3)$ with itself contains the adjoint representation exactly once, namely in the symmetric part. The latter property implies that this case does not lead to a simple Lie superalgebra.

4,5) The tensor product of $\rho(\lambda_1)$ with $\rho(\lambda_3)$ or with $\rho(\lambda_5)$ does not contain the adjoint representation; hence these cases do not lead to a simple Lie superalgebra.

Case C_n , $n \geq 2$

There exists just one admissible pair of representations; the corresponding highest weights and the rank n are

$$\lambda_2 \ , \ \lambda_2 \quad ; \quad n = 3 \ .$$

The tensor product of $\rho(\lambda_2)$ with itself contains the adjoint representation exactly once, namely in the skew-symmetric part. It is well-known that $\rho(\lambda_2)$ is a subrepresentation of $\overset{2}{\Lambda}\rho(\lambda_1)$; thus it is straightforward to construct the candidate for the product mapping $L_{\bar{1}} \times L_{\bar{1}} \longrightarrow L_{\bar{0}}$. Once this has been done it is easy to see that the Jacobi identity for three odd elements is not satisfied.

Cases B_n , $n \geq 3$; D_m , $m \geq 4$; E_6 , E_7 , E_8 , F_4 , G_2

In these cases no admissible pairs of representations do exist.

This concludes the proof of theorem 1 .

5. Extension of some classical simple Z - graded Lie superalgebras

The following proposition is due to Kac [3].

Proposition 3

We suppose that the field K is algebraically closed.

Let $G = \bigoplus_{i \geq -1} G_i$ be a transitive consistently Z - graded Lie superalgebra. Suppose that $G' = G_{-1} \oplus G_0 \oplus G_1$ is a subalgebra of G which is, as a Z - graded Lie superalgebra, isomorphic to one of the following Z - graded Lie superalgebras :

a) spl(n,m) with $n , m \geq 1$; $n \neq m$

   $spl(n,n) / K \cdot I_{2n}$ with $n \geq 3$

   osp(2,2n) with $n \geq 1$

b) b(n + 1) or b'(n + 1) with $n \geq 2$.

It then follows that G = G', i.e. we have $G_i = \{0\}$ if $i \geq 2$.

Remarks

3) For the case of the algebra $spl(2,2)/K\cdot I_4$ we refer the reader to proposition 8 of the next paragraph (the algebra $spl(2,2)/K\cdot I_4$ is isomorphic to $H(4)$).

4) Recall that the Z - graded Lie superalgebra $b'(n+1)$ is obtained from $b(n+1)$ by an inversion of the Z - gradation.

Proof

Since the algebra G is transitive it is sufficient to prove that G_2 is equal to $\{0\}$. We choose a Cartan subalgebra h of G_0. The argument will crucially depend on the properties of the root space decomposition of the classical simple Lie superalgebra G' (see §4, n°6, proposition 1).

a) In this case we can apply to G' the notation and results of §3, n°2.

Suppose that $G_2 \neq \{0\}$. According to §1, proposition 3 the representation of G_0 in G_2 is completely reducible. Let λ be a weight of this representation and let $T_\lambda \in G_2$ be a weight vector associated with λ. Since G is transitive there exists an odd root $-\beta$ of G' and a root vector $X_{-\beta} \in G_{-1}$ associated with $-\beta$ such that $\langle X_{-\beta}, T_\lambda \rangle \neq 0$. Thus $\lambda - \beta$ is an odd root of G' and

$$Y_{\lambda-\beta} = \langle X_{-\beta}, T_\lambda \rangle \in G_1 \tag{5.81}$$

is a root vector associated with it. We know that 0 is not an odd root of G' (see §4, proposition 1), hence we have $\lambda - \beta \neq 0$.

Let $X_{-\lambda+\beta} \in G_{-1}$ be the root vector associated with the odd root $-\lambda+\beta$ of G' which is normalized such that

$$\langle X_{-\lambda+\beta}, Y_{\lambda-\beta} \rangle = H_{\lambda-\beta} \tag{5.82}$$

(see §4, proposition 1 and equation (3.37)). It follows that

$$\langle X_{-\beta}, \langle X_{-\lambda+\beta}, T_\lambda \rangle\rangle = -H_{\lambda-\beta} \,. \tag{5.83}$$

This equation shows that $\langle X_{-\lambda+\beta}, T_\lambda \rangle \in G_1$ is non-zero, hence it is a root vector of G' which is associated with the odd root β of G'. But then the equations (5.83) and (3.37) imply that

$$H_{\lambda-\beta} = r\,H_\beta \tag{5.84}$$

with some element $r \in K$. Hence $\lambda - \beta$ and β are two proportional odd roots of G'. According to proposition 1 of §4 we conclude that

$$r = \pm 1 \; . \tag{5.85}$$

Suppose that $r = -1$ and hence that $\lambda = 0$. Then the two root vectors $X_{-\beta} \in G_{-1}$ and $Y_{\lambda-\beta} \in G_1$ are both associated with the odd root $-\beta$ of G' ; since these vectors are linearly independent, this contradicts proposition 1 of §4 .

It follows that $r = 1$, i.e. that

$$\lambda = 2\beta \; . \tag{5.86}$$

Thus we have shown that every weight of the representation of G_0 in G_2 is equal to twice a weight of the representation of G_0 in G_1 . Since zero is not a weight of the representation of $\langle G_0 , G_0 \rangle$ in G_1 this is impossible.

b) In this case we shall use for $G_0 \simeq A_n$ the notation introduced in the appendix. In particular, we choose a fundamental system of simple roots $\alpha_1, \ldots\ldots, \alpha_n$ and construct the corresponding system $\lambda_1, \ldots\ldots, \lambda_n$ of fundamental weights. (Our enumeration of the vertices of the Dynkin diagram has been specified in the appendix, n°3 .)

We have to treat two cases :

Case 1 : G' is isomorphic to $b(n+1)$; we may assume that the representation of G_0 in G_{-1} (resp. in G_1) is equal to $\rho(2\lambda_1)$ (resp. to $\rho(\lambda_{n-1})$).

Case 2 : G' is isomorphic to $b'(n+1)$; we may assume that the representation of G_0 in G_{-1} (resp. in G_1) is equal to $\rho(\lambda_2)$ (resp. to $\rho(2\lambda_n)$).

We shall discuss both cases simultaneously.

To begin with let us consider the linear mapping

$$f : G_2 \longrightarrow \mathrm{Hom}_K(G_{-1}, G_1) \tag{5.87,a}$$

which is defined by

$$(f(T))(X) = \langle T, X \rangle \quad \text{for all } T \in G_2,\ X \in G_{-1}. \tag{5.87,b}$$

The transitivity of G implies that f is injective, moreover, f is G_0-invariant. Hence the G_0-module G_2 is isomorphic to a G_0-submodule of $\mathrm{Hom}_K(G_{-1}, G_1)$, i.e. to a G_0-submodule of $G_{-1}^* \otimes G_1$.

Let λ (resp. μ) be the highest (resp. lowest) weight of the G_0-module G_{-1} (resp. G_1) and let $X_\lambda \in G_{-1}$ (resp. $Y_\mu \in G_1$) be a weight vector associated with it. We have in the first case

$$\lambda = 2\lambda_1 \quad , \quad \mu = -\lambda_2 \tag{5.88,a}$$

$$-\lambda + \mu = -2\lambda_1 - \lambda_2 \quad , \quad \lambda + \mu = \alpha_1 \tag{5.88,b}$$

and in the second case

$$\lambda = \lambda_2 \quad , \quad \mu = -2\lambda_1 \tag{5.89,a}$$

$$-\lambda + \mu = -2\lambda_1 - \lambda_2 \quad , \quad \lambda + \mu = -\alpha_1 . \tag{5.89,b}$$

Recall that in the first / second case $\langle X_\lambda, Y_\mu \rangle$ is a (non-zero!) root vector of G_0 associated with the root $\pm\alpha_1$ (see §1, proposition 4).

In the following we shall assume that for every root α of G_0 a root vector E_α has been chosen in such a way that

$$\langle E_{-\alpha}, E_\alpha \rangle = H_\alpha \quad \text{for all roots } \alpha \tag{5.90}$$

(see the appendix; the element $H_\alpha \in h$ is defined via some non-degenerate invariant bilinear form on G_0). Moreover, we shall assume that in the first / second case we have

$$\langle X_\lambda, Y_\mu \rangle = E_{\pm\alpha_1} . \tag{5.91}$$

Next we remark that in both cases the representation of G_0 in $G_{-1}^* \otimes G_1$ is equivalent to

$$\rho(2\lambda_n) \otimes \rho(\lambda_{n-1}) \simeq \rho(2\lambda_n + \lambda_{n-1}) \oplus \rho(2\lambda_n + \lambda_{n-1} - \alpha_n - \alpha_{n-1}) . \tag{5.92}$$

The lowest weight of $\rho(2\lambda_n + \lambda_{n-1})$ is equal to

$$\sigma = -2\lambda_1 - \lambda_2 = -\lambda + \mu , \tag{5.93}$$

the lowest weight of $\rho(2\lambda_n + \lambda_{n-1} - \alpha_n - \alpha_{n-1})$ is equal to

$$\tau = -2\lambda_1 - \lambda_2 + \alpha_1 + \alpha_2 = -\lambda + \mu + \alpha_1 + \alpha_2 . \qquad (5.94)$$

We shall now show that none of the two representations on the right hand side of equation (5.92) can be contained in the G_0 - module G_2. It follows that $G_2 = \{0\}$, as required.

Suppose first that the representation $\rho(2\lambda_n + \lambda_{n-1})$ is contained in the G_0 - module G_2. Let $T_\sigma \in G_2$ be a weight vector associated with the lowest weight σ of this subrepresentation. The transitivity of G implies that $\langle X_\lambda , T_\sigma \rangle \neq 0$. Hence $\langle X_\lambda , T_\sigma \rangle \in G_1$ is a weight vector which belongs to the weight μ. But the weights of the representation $\rho(\lambda_{n-1})$ (resp. $\rho(2\lambda_n)$) are simple, thus $\langle X_\lambda , T_\sigma \rangle$ is proportional to Y_μ. Therefore, by multiplying T_σ by a suitable non-zero element of K, we may assume that

$$\langle X_\lambda , T_\sigma \rangle = Y_\mu . \qquad (5.95)$$

This implies that

$$2 E_{\pm\alpha_1} = 2\langle X_\lambda , Y_\mu \rangle = \langle\langle X_\lambda , X_\lambda \rangle , T_\sigma \rangle = 0 , \qquad (5.96)$$

a contradiction.

Suppose next that the representation $\rho(2\lambda_n + \lambda_{n-1} - \alpha_n - \alpha_{n-1})$ is contained in the G_0 - module G_2. Let $T_\tau \in G_2$ be a weight vector associated with the lowest weight τ of this subrepresentation. The transitivity of G implies that $\langle X_\lambda , T_\tau \rangle \neq 0$. Hence $\langle X_\lambda , T_\tau \rangle \in G_1$ is a weight vector which belongs to the weight $\mu + \alpha_1 + \alpha_2$. Consequently, $\langle X_\lambda , T_\tau \rangle$ is proportional to the non-zero element $\langle\langle E_{\alpha_1} , E_{\alpha_2} \rangle , Y_\mu \rangle$ of G_1. Therefore, by multiplying T_τ by a suitable non-zero element of K, we may assume that

$$\langle X_\lambda , T_\tau \rangle = \langle\langle E_{\alpha_1} , E_{\alpha_2} \rangle , Y_\mu \rangle . \qquad (5.97)$$

In the second case the equations (5.91) and (5.97) imply that

$$0 = \langle\langle X_\lambda , X_\lambda \rangle , T_\tau \rangle = -2(\alpha_1 | \alpha_2) E_{\alpha_2} , \qquad (5.98)$$

a contradiction.

In the first case we deduce from equation (5.97) that

$$\langle\langle E_{-\alpha_1}, X_\lambda\rangle, T_\tau\rangle = (\alpha_1|\alpha_2)\langle E_{\alpha_2}, Y_\mu\rangle \, . \tag{5.99}$$

Using the equations (5.91) and (5.99) it is easy to check that

$$0 = \langle\langle\langle E_{-\alpha_1}, X_\lambda\rangle, \langle E_{-\alpha_1}, X_\lambda\rangle\rangle, T_\tau\rangle = -2(\alpha_1|\alpha_2)^2 E_{\alpha_2} \, . \tag{5.100}$$

Again this is a contradiction.

§ 6 The Cartan Lie superalgebras [3]

In the present paragraph we shall construct four additional sequences of simple Lie superalgebras which, if the field K is algebraically closed, complete the list of all simple Lie superalgebras. Throughout the whole paragraph V will denote a *finite-dimensional* vector space over the (arbitrary) field K .

1. The Lie superalgebra W(V) of superderivations of an exterior algebra

A. Definition and elementary properties of W(V)

Let V be a finite-dimensional vector space; we set

$$\dim V = n \quad . \tag{6.1}$$

Recall that in chapter I, §1, example 4) we have defined, for an arbitrary superalgebra T , the Lie superalgebra $\mathfrak{D}(T)$ of superderivations of T . It is well-known that the exterior algebra $\Lambda V = \bigoplus_{s=0}^{n} \overset{s}{\Lambda} V$ is an associative Z - graded algebra. The Z - gradation induces a Z_2 - gradation on ΛV (see chapter 0, §2, 1)); thus ΛV may be considered as a superalgebra and hence the Lie superalgebra of superderivations of ΛV is well-defined. Adopting the notation introduced by Kac [3] we shall write W(V) instead of $\mathfrak{D}(\Lambda V)$.

Let D be any element of W(V) . Obviously, we have

$$D(1) = 0 \quad . \tag{6.2}$$

Since the algebra ΛV is generated by 1 and V we conclude that D is uniquely fixed once we are given the restriction of D to V . Conversely, it is easy to see that for every linear mapping

$$A : V \longrightarrow \Lambda V \tag{6.3}$$

there exists a (unique) superderivation D_A of ΛV which extends A . It follows that

$$\dim W(V) = n \cdot 2^n \quad . \tag{6.4}$$

The algebra $W(V)$ has a natural Z - gradation. Let $r \in Z$; we define

$$W_r(V) = \{ D \in W(V) \mid D(V) \subset \overset{r+1}{\Lambda} V \} . \tag{6.5}$$

Obviously, this definition implies that

$$D(\overset{s}{\Lambda} V) \subset \overset{r+s}{\Lambda} V \quad \text{for all } r, s \in Z \text{ and all } D \in W_r(V) . \tag{6.6}$$

It follows that $(W_r(V))_{r \in Z}$ is a Z - gradation of the algebra $W(V)$ which is consistent with the Z_2 - gradation. We have

$$W_r(V) = \{0\} \quad \text{if } r \leqslant -2 \text{ or } r \geqslant n \tag{6.7}$$

and

$$\dim W_r(V) = n\binom{n}{r+1} \quad \text{if } -1 \leqslant r \leqslant n-1 . \tag{6.8}$$

The Z - graded algebra $W(V)$ may be considered as a left Z - graded ΛV - module : Let $r, s \in Z$ and let $a \in \overset{s}{\Lambda} V$, $D \in W_r(V)$; then $a \wedge D$ is a superderivation of ΛV and we have

$$a \wedge D \in W_{r+s}(V) . \tag{6.9}$$

The special cases $r = -1$ and $r = 0$ of equation (6.5) are particularly important. By definition, $W_{-1}(V)$ consists of the antiderivations D_g with $g \in V^*$; moreover, we have

$$D_g(x_1 \wedge \ldots\ldots \wedge x_s) = \sum_{i=1}^{s} (-1)^{i+1} g(x_i)\, x_1 \wedge \ldots \wedge \hat{x}_i \wedge \ldots \wedge x_s \tag{6.10}$$

$$\text{for all } s \geqslant 1 \text{ and all } x_1, \ldots\ldots, x_s \in V$$

(where the hat ^ indicates that the factor x_i has to be deleted). From equation (6.7) we conclude that

$$D_f \circ D_g = - D_g \circ D_f \quad \text{for all } f, g \in V^* ; \tag{6.11}$$

of course, this relation may also be checked using the explicit formula (6.10) .

On the other hand, $W_0(V)$ consists of the derivations D_A with $A \in \mathrm{Hom}(V)$; we have

$$D_A(x_1 \wedge \ldots\ldots \wedge x_s) = \sum_{i=1}^{s} x_1 \wedge \ldots \wedge (Ax_i) \wedge \ldots \wedge x_s \qquad (6.12)$$

for all $s \geqslant 1$ and all $x_1, \ldots\ldots, x_s \in V$.

Thus $A \longrightarrow D_A$ is just the canonical representation of $gl(V)$ in ΛV ; in particular, it follows that

$$\langle D_A , D_B \rangle = D_{[A,B]} \quad \text{for all } A , B \in gl(V) . \qquad (6.13)$$

Consequently, the Lie algebra $W_0(V)$ may be identified with the general linear Lie algebra $gl(V)$ of V .

Let id be the identity mapping of V onto itself. Then

$$C = D_{id} \qquad (6.14)$$

generates the center of $W_0(V)$ and we have

$$C(a) = s\,a \quad \text{for all } a \in \overset{s}{\Lambda} V \text{ and all } s \in Z \qquad (6.15)$$

as well as

$$\langle C , D \rangle = r\,D \quad \text{for all } D \in W_r(V) \text{ and all } r \in Z . \qquad (6.16)$$

Next we shall consider a realization of the algebra $W(V)$ which more clearly exhibits the tensor character of the elements of $W_r(V)$. Let us define a linear mapping

$$\omega : (\Lambda V) \otimes V^* \longrightarrow W(V) \qquad (6.17,a)$$

by the requirement that

$$\omega(a \otimes g) = a \wedge D_g \quad \text{for all } a \in \Lambda V , g \in V^* . \qquad (6.17,b)$$

Moreover, let us consider V^* as a Z - graded vector space all of whose elements have a degree equal to -1 . Then the tensor product $(\Lambda V) \otimes V^*$ has a natural Z - gradation and it is easy to see that ω is an isomorphism of Z - graded vector spaces.

Thus the mapping ω may be used to transport the algebra structure of $W(V)$ to $(\Lambda V) \otimes V^*$. The multiplication in $(\Lambda V) \otimes V^*$ is then given by

$$\langle a \otimes f , b \otimes g \rangle = (a \wedge (D_f b)) \otimes g + (-1)^s ((D_g a) \wedge b) \otimes f \qquad (6.18)$$

$$\text{for all } a \in \overset{s}{\Lambda} V , \; b \in \Lambda V \; ; \; f , g \in V^* .$$

Note that ω induces a linear mapping $V \otimes V^* \longrightarrow W_0(V)$. If we identify the elements of $W_0(V)$ with their restriction to V this mapping is just the canonical isomorphism of $V \otimes V^*$ onto $\mathrm{Hom}(V)$.

The equation (6.18) implies that

$$\langle \omega^{-1}(D_A) , b \otimes g \rangle = (D_A b) \otimes g + b \otimes (-{}^t A g) \qquad (6.19)$$

$$\text{for all } A \in gl(V) , \; b \in \Lambda V , \; g \in V^* ,$$

hence $A \longrightarrow \langle \omega^{-1}(D_A) , \cdot \rangle$ is the canonical representation of the Lie algebra $gl(V)$ in the tensor space $(\Lambda V) \otimes V^*$.

Of course, the Z-graded Lie superalgebras $W(V)$ and $(\Lambda V) \otimes V^*$ may be identified by means of the mapping ω; it will depend on the circumstances which of these two realizations is more advantageous. (A third description of this algebra will be given in subsection C.)

It is now easy to prove the following proposition:

Proposition 1

Let V be an n-dimensional vector space.

a) The Z-graded Lie superalgebra $W(V)$ is transitive.

b) If $n \geq 2$, then

$$\langle W_r(V) , W_1(V) \rangle = W_{r+1}(V) \quad \text{for all } r \geq -1 . \qquad (6.20)$$

c) The Lie superalgebra $W(V)$ is simple provided that $n \geq 2$

Proof

To prove a) one may take advantage of the subsequent lemma, the statement b) is easily checked directly, finally, c) follows from lemma 4 of §2.

Lemma 1

Let a be a non-zero element of $\overset{s}{\Lambda} V$, $s \geq 1$. Then there exist linear forms $g \in V^*$ such that $D_g(a) \neq 0$.

Remark 1)

Let $A : V \longrightarrow V'$ be an isomorphism of the vector space V onto a second vector space V'. It is well-known that there exists a unique isomorphism $\hat{A} : \Lambda V \longrightarrow \Lambda V'$ of Z - graded algebras which extends A . Obviously, the isomorphism $\hat{A}$ induces an isomorphism $W(V) \longrightarrow W(V')$ of Z - graded Lie superalgebras which is given by

$$D \longrightarrow \hat{A} \circ D \circ \hat{A}^{-1} \quad \text{if } D \in W(V) \; ; \tag{6.21}$$

moreover, the corresponding isomorphism of $(\Lambda V) \otimes V^*$ onto $(\Lambda V') \otimes V'^*$ is equal to $\hat{A} \otimes {}^t A^{-1}$.

In particular, the Z - graded Lie superalgebra W(V) is isomorphic to $W(K^n)$; in the following the latter algebra will be denoted by W(n) .

B. W(V) as a sl(V) - module

According to part A of this section the Lie algebra $W_0(V)$ is canonically isomorphic to gl(V) , hence W(V) has a natural structure of a gl(V) - module. Since the action on W(V) of the center of gl(V) is already known (see equation (6.16)) we may restrict our attention to the sl(V) - module structure of W(V) . In the following we shall assume that

$$\dim V = n \geq 2 \; . \tag{6.22}$$

Let $e_1, \ldots\ldots, e_n$ be a basis of the vector space V and let h be the subspace of sl(V) consisting of those elements whose matrices with respect to the basis $e_1, \ldots\ldots, e_n$ are diagonal. It is well-known that h is a splitting Cartan subalgebra [33] of sl(V) . Thus we may use part of the notation of the appendix even if the field K is not algebraically closed.

We choose a fundamental system of simple roots of sl(V) with respect to h and construct the corresponding system $\lambda_1, \ldots\ldots, \lambda_{n-1}$ of fundamental weights. We assume that our choice is such that λ_s , $1 \leq s \leq n-1$, is the highest weight of the canonical representation of sl(V) in $\overset{s}{\Lambda} V$. For convenience, we define

$$\lambda_0 = \lambda_n = 0 \ . \tag{6.23}$$

According to part A the representation of sl(V) in $W_r(V)$, $-1 \leq r \leq n-1$, is equivalent to $\rho(\lambda_{r+1}) \otimes \rho(\lambda_{n-1})$. For $r = -1$ and $r = n-1$ this representation is equivalent to $\rho(\lambda_{n-1})$. On the other hand we have

$$\rho(\lambda_{r+1}) \otimes \rho(\lambda_{n-1}) \simeq \rho(\lambda_{r+1} + \lambda_{n-1}) \oplus \rho(\lambda_r) \quad \text{if } 0 \leq r \leq n-2 \ . \tag{6.24}$$

We are going to determine the two subspaces of $W_r(V)$ which correspond to the two irreducible representations on the right hand side of equation (6.24) . The discussion is most easily carried out in the algebra $(\Lambda V) \otimes V^*$. Let

$$\tau : (\Lambda V) \otimes V^* \longrightarrow \Lambda V \tag{6.25,a}$$

be the linear mapping which is defined by

$$\tau(a \otimes g) = D_g(a) \quad \text{for all } a \in \Lambda V , \ g \in V^* \ . \tag{6.25,b}$$

As in the previous subsection we consider V^* as a Z - graded vector space all of whose elements have a degree equal to -1 . Then the mapping τ is homogeneous of degree 0 .

On the other hand, let $Sym_2(V^*)$ denote the subspace of the symmetric algebra of V^* consisting of those elements which (according to the usual definition) are homogeneous of degree 2 [34]. For our purposes it is advantageous to attach to the elements of $Sym_2(V^*)$ the degree -2 . The canonical mapping $V^* \times V^* \longrightarrow Sym_2(V^*)$ will be written as $(f,g) \longrightarrow f \cdot g$, with $f , g \in V^*$.

We define a linear mapping

$$\sigma : (\Lambda V) \otimes Sym_2(V^*) \longrightarrow (\Lambda V) \otimes V^* \tag{6.26,a}$$

by the requirement that

$$\sigma(a \otimes (f\cdot g)) \;=\; (D_f a) \otimes g \;+\; (D_g a) \otimes f \qquad (6.26,b)$$

for all $a \in \Lambda V$ and $f, g \in V^*$.

According to our conventions the mapping σ is homogeneous of degree 0 .

It is easy to see that τ and σ are homomorphisms of $gl(V)$ - modules, furthermore, the sequence

$$(\Lambda V) \otimes Sym_2(V^*) \xrightarrow{\ \sigma\ } (\Lambda V) \otimes V^* \xrightarrow{\ \tau\ } \Lambda V \qquad (6.27)$$

is exact.

More precisely, for any $r \in Z$, let

$$(\overset{r+2}{\Lambda} V) \otimes Sym_2(V^*) \xrightarrow{\ \sigma_r\ } (\overset{r+1}{\Lambda} V) \otimes V^* \xrightarrow{\ \tau_r\ } \overset{r}{\Lambda} V \qquad (6.28)$$

denote the sequence of $gl(V)$ - modules which is induced by (6.27) . Then we have :

a) Let $-1 \leqslant r \leqslant n-2$. Then $\text{kernel}(\tau_r) = \text{image}(\sigma_r)$ is the subspace of $(\overset{r+1}{\Lambda} V) \otimes V^*$ which belongs to the representation $\rho(\lambda_{r+1} + \lambda_{n-1})$. The corresponding subspace of $W_r(V)$ will be denoted by $S_r(V)$. We extend this definition by setting

$$S_r(V) \;=\; \{0\} \quad \text{if } r \leqslant -2 \text{ or } r \geqslant n-1 \; . \qquad (6.29)$$

b) Let $0 \leqslant r \leqslant n-1$. Then the subspace of $(\overset{r+1}{\Lambda} V) \otimes V^*$ which belongs to $\rho(\lambda_r)$ is bijectively mapped by τ_r onto $\overset{r}{\Lambda} V$. The subspace of $W_r(V)$ corresponding to the representation $\rho(\lambda_r)$ will be denoted by $T_r(V)$. We extend this definition by setting

$$T_r(V) \;=\; \{0\} \quad \text{if } r \leqslant -1 \text{ or } r \geqslant n \; . \qquad (6.30)$$

It is easy to see that

$$S(V) \;=\; \bigoplus_{r=-1}^{n-2} S_r(V) \;=\; \text{kernel}(\tau \circ \omega^{-1}) \;=\; \text{image}(\omega \circ \sigma) \qquad (6.31)$$

is a subalgebra of $W(V)$; this algebra will be discussed in the next section.

The subspace $T_r(V)$ of $W_r(V)$ may be described as follows. Consider the linear mapping

$$\alpha : \Lambda V \longrightarrow W(V) \tag{6.32,a}$$

which is defined by

$$\alpha(a) = a \wedge C \quad \text{if } a \in \Lambda V . \tag{6.32,b}$$

Here C is the element of the center of $W_0(V)$ which has been defined in equation (6.14) . Obviously, α is homogeneous of degree 0 , moreover, α is a homomorphism of gl(V) - modules. Finally, it is not difficult to check that

$$\tau \circ \omega^{-1} \circ \alpha(a) = (-1)^r (n-r)\, a \quad \text{for all } a \in \overset{r}{\Lambda} V \text{ and all } r \in Z . \tag{6.33}$$

Hence, for $r \neq n$, the mapping α induces an isomorphism $\overset{r}{\Lambda} V \longrightarrow T_r(V)$ of gl(V) - modules. In particular, we have

$$T_r(V) = (\overset{r}{\Lambda} V) \wedge C \quad \text{for all } r \in Z . \tag{6.34}$$

Obviously, our definitions imply that

$$W_r(V) = S_r(V) \oplus T_r(V) \quad \text{for all } r \in Z . \tag{6.35}$$

Suppose that $-1 \leq r \leq n-1$ and let $b \in \overset{r+1}{\Lambda} V$, $g \in V^*$. Then we deduce from equation (6.33) that the component of $b \wedge D_g$ in $T_r(V)$ (with respect to the decomposition (6.35)) is equal to

$$\frac{(-1)^r}{n-r} D_g(b) \wedge C . \tag{6.36}$$

This remark will be used to prove the following lemma.

Lemma 2

Let $1 \leq r \leq n-1$ and let D be a non-zero element of $T_r(V)$. Then there exists a linear form $g \in V^*$ such that the component of $\langle D_g , D \rangle$ in $T_{r-1}(V)$ is different from zero.

Proof

In view of equation (6.34) we have

$$D = a \wedge C \tag{6.37}$$

with some non-zero element $a \in \overset{r}{\bigwedge} V$. Let $g \in V^*$; using the above remark it follows that the component of $\langle D_g, a \wedge C \rangle$ in $T_{r-1}(V)$ is equal to

$$\frac{n-r}{n-r+1} D_g(a) \wedge C . \tag{6.38}$$

Our assertion now follows from lemma 1.

C. W(V) as a universal transitive Z - graded Lie superalgebra

To begin with we shall describe a third realization of the algebra W(V). For convenience, we actually consider the algebra $W(V^*)$ instead of W(V).

In part A of this section we have introduced the isomorphism

$$\omega : (\bigwedge V^*) \otimes V \longrightarrow W(V^*) \tag{6.39}$$

of Z - graded Lie superalgebras. It is well-known that the vector space $\bigwedge V^*$ is canonically isomorphic to $(\bigwedge V)^*$; moreover, the vector space $(\bigwedge V)^* \otimes V$ is canonically isomorphic to $\mathrm{Hom}(\bigwedge V, V)$. We use these isomorphisms to transport the Z - gradation and the algebra structure from $(\bigwedge V^*) \otimes V$ to $\mathrm{Hom}(\bigwedge V, V)$.

Without going into the details we shall give the result of this construction. A linear mapping $\bigwedge V \longrightarrow V$ is homogeneous of degree $r \in Z$ if and only if its restriction to $\overset{s}{\bigwedge} V$ is equal to zero for all $s \neq r+1$. Consequently, $\mathrm{Hom}(\bigwedge V, V)_r$ may be identified with $\mathrm{Hom}(\overset{r+1}{\bigwedge} V, V)$.

Let $-1 \leqslant r \leqslant n-1$. The canonical mapping

$$\Omega_r : (\overset{r+1}{\bigwedge} V^*) \otimes V \longrightarrow \mathrm{Hom}(\overset{r+1}{\bigwedge} V, V) \tag{6.40}$$

is defined as follows.

In the case $r = -1$ we require that

$$(\Omega_{-1}(1 \otimes x))(1) = x \quad \text{for all } x \in V . \tag{6.41}$$

If $0 \leq r \leq n-1$, we define

$$(\Omega_r((y_1' \wedge \dots \wedge y_{r+1}') \otimes x))(x_1 \wedge \dots \wedge x_{r+1}) = \det(y_i'(x_j))\, x \qquad (6.42)$$

for all $y_1', \dots, y_{r+1}' \in V^*$ and $x, x_1, \dots, x_{r+1} \in V$.

On the identification mentioned above the family $(\Omega_r)_{-1 \leq r \leq n-1}$ defines an isomorphism

$$\Omega : (\wedge V^*) \otimes V \longrightarrow \mathrm{Hom}(\wedge V, V) \qquad (6.43)$$

of Z-graded vector spaces.

The multiplication in $\mathrm{Hom}(\wedge V, V)$ is given by the following prescription.

Let $r, s \geq 0$ and let

$$A \in \mathrm{Hom}(\overset{r+1}{\wedge} V, V) \quad , \quad B \in \mathrm{Hom}(\overset{s+1}{\wedge} V, V) . \qquad (6.44)$$

Then $\langle A, B \rangle$ is the element of $\mathrm{Hom}(\overset{r+s+1}{\wedge} V, V)$ which satisfies

$$\langle A, B \rangle (x_1 \wedge \dots \wedge x_{r+s+1}) \qquad (6.45)$$

$$= \sum_{\pi \in H_{r+1,s}} \varepsilon(\pi)\, B((A(x_{\pi(1)} \wedge \dots \wedge x_{\pi(r+1)})) \wedge x_{\pi(r+2)} \wedge \dots \wedge x_{\pi(r+s+1)})$$

$$- (-1)^{rs} \sum_{\pi \in H_{s+1,r}} \varepsilon(\pi)\, A((B(x_{\pi(1)} \wedge \dots \wedge x_{\pi(s+1)})) \wedge x_{\pi(s+2)} \wedge \dots \wedge x_{\pi(r+s+1)})$$

for all $x_1, \dots, x_{r+s+1} \in V$.

Here $H_{r+1,s}$ denotes the set of all permutations of $\{1,2,\dots, r+s+1\}$ which are increasing on the subsets $\{1,\dots, r+1\}$ and $\{r+2,\dots, r+s+1\}$; the set $H_{s+1,r}$ is defined similarly. Finally, $\varepsilon(\pi)$ is the signum of the permutation π. (If $s = 0$, we set $x_{\pi(r+2)} \wedge \dots \wedge x_{\pi(r+s+1)}$ equal to 1; similarly for $r = 0$.)

If adequately interpreted, the equation (6.45) remains valid if either $r = -1$ or $s = -1$. As an example we consider the case $r = -1$, $s \geq 0$. Let

$$A \in \mathrm{Hom}(\overset{0}{\wedge} V, V) \quad , \quad B \in \mathrm{Hom}(\overset{s+1}{\wedge} V, V) . \qquad (6.46)$$

Then

$$\langle A, B\rangle(x_1 \wedge \ldots \wedge x_s) = B((A(1)) \wedge x_1 \wedge \ldots \wedge x_s) \qquad (6.47)$$
$$\text{for all } x_1, \ldots, x_s \in V .$$

(Again, we set $x_1 \wedge \ldots \wedge x_s$ equal to 1 if $s = 0$.)

Of course, we have $\langle A, B\rangle = 0$ if both A and B are homogeneous of degree -1.

Now let $G = \bigoplus_{r \geq -1} G_r$ be a transitive consistently Z-graded Lie superalgebra such that G_{-1} is finite-dimensional. Let $r \geq 0$ be an integer and let $Y \in G_r$. Consider the $r+1$-linear mapping

$$\underbrace{G_{-1} \times G_{-1} \times \ldots \times G_{-1}}_{r+1 \text{ factors}} \longrightarrow G_{-1} \qquad (6.48,a)$$

which is defined by

$$(X_1, X_2, \ldots, X_{r+1}) \longrightarrow \langle X_1, \langle X_2, \ldots, \langle X_{r+1}, Y\rangle ..\rangle\rangle \qquad (6.48,b)$$
$$\text{for all } X_1, \ldots, X_{r+1} \in G_{-1} .$$

Since $\langle G_{-1}, G_{-1}\rangle = \{0\}$ this mapping is totally skew-symmetric. Hence there exists a unique linear mapping

$$\mu_r(Y) : \bigwedge^{r+1} G_{-1} \longrightarrow G_{-1} \qquad (6.49,a)$$

such that

$$(\mu_r(Y))(X_1 \wedge \ldots \wedge X_{r+1}) = \langle X_1, \ldots, \langle X_{r+1}, Y\rangle ..\rangle \qquad (6.49,b)$$
$$\text{for all } X_1, \ldots, X_{r+1} \in G_{-1} .$$

Thus we have defined, for all $r \geq 0$, a linear mapping

$$\mu_r : G_r \longrightarrow \operatorname{Hom}(\bigwedge^{r+1} G_{-1}, G_{-1}) . \qquad (6.50)$$

Moreover, let

$$\mu_{-1} : G_{-1} \longrightarrow \operatorname{Hom}(\bigwedge^{0} G_{-1}, G_{-1}) \qquad (6.51,a)$$

be the canonical isomorphism, i.e.

$$(\mu_{-1}(Y))(1) = Y \quad \text{for all } Y \in G_{-1} . \tag{6.51,b}$$

Finally, if $r \in Z$, $r \leqslant -2$, we define μ_r to be the zero mapping of G_r into $\mathrm{Hom}(\overset{r+1}{\bigwedge} G_{-1}, G_{-1})$.

Obviously, the transitivity of G implies that all of the mappings μ_r, $r \in Z$, are injective. Furthermore, it is not difficult to check that

$$\mu_{r+s}(\langle X , Y\rangle) = (-1)^{rs} \langle \mu_r(X) , \mu_s(Y)\rangle \tag{6.52}$$

for all $X \in G_r$, $Y \in G_s$ and all $r , s \in Z$.

(Of course, the product on the right hand side of equation (6.52) has to be calculated according to equation (6.45) .)

Let us define, for all $r \in Z$, the linear mapping

$$\tilde{\mu}_r : G_r \longrightarrow \mathrm{Hom}(\overset{r+1}{\bigwedge} G_{-1} , G_{-1}) \tag{6.53,a}$$

by

$$\tilde{\mu}_r = (-1)^{\frac{1}{2}r(r+1)} \mu_r , \tag{6.53,b}$$

moreover, let

$$\tilde{\mu} : G \longrightarrow \mathrm{Hom}(\bigwedge G_{-1} , G_{-1}) \tag{6.54}$$

be the injective linear mapping defined by the family $(\tilde{\mu}_r)_{r\in Z}$. Then the equation (6.52) implies that $\tilde{\mu}$ is a homomorphism of Z - graded Lie superalgebras. Thus we have shown :

<u>Theorem</u> 1

Let $G = \bigoplus_{r \geqslant -1} G_r$ be a transitive consistently Z - graded Lie superalgebra such that G_{-1} is finite-dimensional. Then $\omega \circ \Omega^{-1} \circ \tilde{\mu} : G \longrightarrow W(G^*_{-1})$ is a (canonical) *injective* homomorphism of Z - graded Lie superalgebras. Moreover, $\omega \circ \Omega^{-1} \circ \tilde{\mu}$ induces the canonical vector space isomorphism of G_{-1} onto $W_{-1}(G^*_{-1})$.

Theorem 1 and proposition 1 imply the following corollary.

Corollary

Let $G = \bigoplus_{r \geq -1} G_r$ be a transitive consistently Z - graded Lie superalgebra such that, for some positive integer $n \geq 2$,

$$\dim G_{-1} = n \quad ; \quad \dim G_1 = n\binom{n}{2} \ . \tag{6.55}$$

Then the Z - graded Lie superalgebras G and W(n) are isomorphic.

D. W(V) as a universal transitive filtered Lie superalgebra

We remind the reader (see §1, $n^0 4$) that every Z - graded Lie superalgebra $G = \bigoplus_{r \geq -1} G_r$ admits a canonical filtration. Accordingly, we define, for every $s \in Z$,

$$W^s(V) = \bigoplus_{r \geq s} W_r(V) \ ; \tag{6.56}$$

the family $(W^s(V))_{s \in Z}$ is called the *canonical filtration of* W(V) . Let us agree that a subalgebra L of W(V) will always be filtered by setting

$$L^s = L \cap W^s(V) \quad \text{for all } s \in Z \ . \tag{6.57}$$

Theorem 2

Let L be a transitive filtered Lie superalgebra, equipped with the filtration $(L^r)_{r \in Z}$ (see §1, $n^0 4$, definition 3). Suppose that $L_{\bar{0}} \subset L^0$ and that the vector space L/L^0 is finite-dimensional. Let V be any vector space such that

$$\dim V = \dim L/L^0 \ . \tag{6.58}$$

a) There exists an isomorphism of the filtered Lie superalgebra L onto a filtered Z_2 - graded subalgebra of W(V) .

b) The subspaces $\bigoplus_{r \geq s} \overset{r}{\Lambda} V$, $s \in Z$, form a (downward) filtration of ΛV .

Any two isomorphisms of the type described in a) are conjugate to each other under an automorphism of the filtered superalgebra ΛV ; this automorphism is uniquely determined.

Proof

Of course, it is sufficient to prove the theorem for the special vector space

$$V = (L/L^0)^* \qquad (6.59)$$

(see remark 1)). In this case our assertion follows from the Guillemin, Sternberg theorem (see chapter I, §4, corollary 1 to theorem 2), applied to the graded subalgebra $L' = L^0$ of L . Let us use the notation introduced in the cited paragraph.

By assumption, L/L^0 is an odd vector space, hence the supersymmetric algebra $\tilde{U}(L/L^0)$ is nothing but the exterior algebra $\Lambda(L/L^0)$. Therefore, the underlying vector space of the algebra $\tilde{F} = \mathrm{Hom}_K(\tilde{U}(L/L^0),K)$ is equal to $(\Lambda(L/L^0))^*$. We are going to show that the algebras $\tilde{F}$ and $\Lambda(L/L^0)^*$ are canonically isomorphic.

Let us define a linear mapping

$$\lambda : \Lambda(L/L^0)^* \longrightarrow (\Lambda(L/L^0))^* \qquad (6.60,a)$$

by the following prescription [19].

First of all, $\lambda(1)$ is the linear form on $\Lambda(L/L^0)$ which vanishes on all of the subspaces $\overset{s}{\Lambda}(L/L^0)$, $s \geqslant 1$, and which satisfies

$$(\lambda(1))(1) = 1 . \qquad (6.60,b)$$

Next let $r \geqslant 1$ and let $y_1', \ldots\ldots, y_r' \in (L/L^0)^*$. Then $\lambda(y_1' \wedge \ldots\ldots \wedge y_r')$ is the linear form on $\Lambda(L/L^0)$ which vanishes on all of the subspaces $\overset{s}{\Lambda}(L/L^0)$ with $s \neq r$ and which satisfies

$$(\lambda(y_1' \wedge \ldots\ldots \wedge y_r'))(x_1 \wedge \ldots\ldots \wedge x_r) = (-1)^{\frac{1}{2}r(r+1)} \det(y_i'(x_j)) \qquad (6.60,c)$$

$$\text{for all } x_1, \ldots\ldots, x_r \in L/L^0 .$$

(Thus, apart from the sign factors, λ is just the usual vector space isomorphism of $\Lambda(L/L^0)^*$ onto $(\Lambda(L/L^0))^*$.)

It is not difficult to see that λ is an isomorphism of the filtered superalgebra $\Lambda(L/L^0)^*$ onto the filtered superalgebra $\tilde{F}$; moreover, λ induces an isomorphism $W((L/L^0)^*) \longrightarrow \tilde{D}$ of filtered Lie superalgebras. This proves part a) of our theorem.

From now on we shall identify $\tilde{F}$ with $\Lambda(L/L^0)^*$ and $\tilde{D}$ with $W((L/L^0)^*)$ by means of the isomorphism λ.

Let θ and η be two isomorphisms of the type described in part a). The mapping $\theta : L \longrightarrow W((L/L^0)^*)$ induces in the obvious way a linear mapping

$$\theta'_{-1} : L \longrightarrow W_{-1}((L/L^0)^*) \ . \tag{6.61}$$

Since $\theta(L^0) \subset W^0((L/L^0)^*)$, this mapping vanishes on L^0. Hence θ'_{-1} induces a linear mapping

$$\theta_{-1} : L/L^0 \longrightarrow W_{-1}((L/L^0)^*) \tag{6.62,a}$$

such that

$$\theta_{-1}(\tilde{A}) = \theta'_{-1}(A) \quad \text{for all } A \in L \ . \tag{6.62,b}$$

(Recall that $\tilde{A}$ denotes the canonical image of A in L/L^0.) The assumptions about θ imply that θ_{-1} is bijective. Applying the same procedure to η instead of θ we obtain the mapping η_{-1}.

Let

$$\gamma : W_{-1}((L/L^0)^*) \longrightarrow L/L^0 \tag{6.63}$$

be the canonical isomorphism of vector spaces. Then

$$\beta = \gamma \circ \eta_{-1} \circ (\theta_{-1})^{-1} \circ \gamma^{-1} \tag{6.64}$$

is an automorphism of the vector space L/L^0. Let ψ be the automorphism of the Z-graded algebra $\Lambda(L/L^0)^*$ which is canonically induced by ${}^t\beta$. Then it is not difficult to prove that

$$\eta(A) - \psi^{-1} \circ \theta(A) \circ \psi \in W^0((L/L^0)^*) \quad \text{for all } A \in L \ . \tag{6.65}$$

Suppose now that η is an isomorphism of the type described in the Guillemin, Sternberg theorem. Then the equation (6.65) shows that the same holds true for the mapping

$$A \longrightarrow \psi^{-1} \circ \theta(A) \circ \psi \quad , \quad A \in L \ . \tag{6.66}$$

Consequently, η and the mapping (6.66) are conjugate under an automorphism of the filtered superalgebra $\Lambda(L/L^0)^*$, and this automorphism is uniquely determined. Evidently, this result implies that b) holds in general.

The isomorphisms of L onto a subalgebra of $W((L/L^0)^*)$ which have been constructed in the proof of theorem 2 suffer from the disadvantage of being non-canonical. Hence we shall now look for suitable additional requirements which might help to convert these isomorphisms into some "normal form". As before, V will denote a finite-dimensional vector space.

Proposition 2

a) If the subalgebra S of the Lie algebra $W_{\bar{0}}(V) = \bigoplus_{r\geq 0} W_{2r}(V)$ is semi-simple, then there exists an automorphism Δ of the filtered Lie super-algebra W(V) such that

$$\Delta(S) \subset S_0(V) . \tag{6.67}$$

b) Let U be a subspace of $W_{\bar{1}}(V) = \bigoplus_{r\geq 0} W_{2r-1}(V)$. Suppose that

$$\langle U, U\rangle = \{0\} \tag{6.68}$$

and that the canonical projection $W(V) \longrightarrow W_{-1}(V)$ induces a bijection of U onto $W_{-1}(V)$. Then there exists an automorphism Δ of the filtered Lie superalgebra W(V) such that

$$\Delta(U) = W_{-1}(V) . \tag{6.69}$$

Proof

a) It is easy to see that

$S_0(V) \simeq sl(V)$ is a Levi factor of $W_{\bar{0}}(V)$,

$T_0(V) \oplus \bigoplus_{r\geq 1} W_{2r}(V)$ is the radical of $W_{\bar{0}}(V)$,

$\bigoplus_{r\geq 1} W_{2r}(V)$ is the nilpotent radical of $W_{\bar{0}}(V)$.

Thus, according to the Levi, Malcev theorem [35], there exists an element

$$X \in \bigoplus_{r\geqslant 1} W_{2r}(V) \tag{6.70}$$

such that the automorphism

$$\Delta = \exp(\mathrm{ad}_W X) \tag{6.71}$$

of the Lie superalgebra $W(V)$ maps S into $S_0(V)$. Obviously, Δ preserves the filtration of $W(V)$.

b) This part follows from theorem 2 itself. In fact, our assumptions imply that U is a filtered Z_2 - graded subalgebra of $W(V)$ such that

$$U_{\bar{0}} = \{0\} \quad , \quad U_{\bar{1}} = U \ , \tag{6.72}$$

the filtration $(U^r)_{r\in Z}$ being given by

$$U^r = U \quad \text{if } r \leqslant -1 \tag{6.73,a}$$

$$U^r = \{0\} \text{ if } r \geqslant 0 \ . \tag{6.73,b}$$

Furthermore, we have

$$\dim U = \dim V \ . \tag{6.74}$$

Of course, the injection of U into $W(V)$ satisfies the conditions of theorem 2,a), moreover, the same holds true for any bijective linear mapping of U onto $W_{-1}(V)$. Our assertion now follows from theorem 2,b).

Remark 2)

Part b) of proposition 2 can be proved more directly without having recourse to theorem 2 . As a preparatory step one derives the following

Lemma 3

Let $r \geqslant 0$ be an integer. Suppose we are given a linear mapping

$$\delta : V^* \longrightarrow W_{r-1}(V) \ . \tag{6.75}$$

Then there exists an element $D \in W_r(V)$ such that

$$\delta(f) = \langle D_f , D \rangle \quad \text{for all } f \in V^* \tag{6.76}$$

if and only if δ satisfies the following condition :

$$\langle D_f , \delta(g)\rangle + \langle D_g , \delta(f)\rangle = 0 \quad \text{for all } f , g \in V^* . \tag{6.77}$$

If this condition is fulfilled the element D is uniquely determined.

(The antiderivations D_g , $g \in V^*$, have been defined in the subsection A above, in particular, see equation (6.10).)

2. The Lie superalgebras $S(V)$ and $\tilde{S}(V,t)$

As before, V denotes an n-dimensional vector space.

A. Elementary properties of $S(V)$

In the preceding section we have defined the Z-graded subspace

$$S(V) = \bigoplus_{r=-1}^{n-2} S_r(V) \tag{6.78}$$

of $W(V)$. Recalling the well-known fact that the vector space $Sym_2(V^*)$ is generated by the elements of the form $g \cdot g$ with $g \in V^*$ we conclude from the definition that the vector space $S_r(V)$ is generated by the superderivations of the form

$$(D_g a) \wedge D_g \quad \text{with } a \in \bigwedge^{r+2} V , \ g \in V^* , \tag{6.79}$$

for all $r \in Z$. It is then easy to check that $S(V)$ is a Z-graded subalgebra of $W(V)$ (a fact which has already been mentioned before).

Evidently

$$S_{-1}(V) = W_{-1}(V) ; \tag{6.80}$$

moreover, the canonical isomorphism $gl(V) \longrightarrow W_0(V)$ induces a canonical isomorphism $sl(V) \longrightarrow S_0(V)$ which will be used to identify the Lie algebra $S_0(V)$ with $sl(V)$.

The representation of $sl(V)$ in $S_r(V)$ has been described in section 1, B ; we conclude that

$$\dim S_r(V) = (n-r-1)\binom{n+1}{r+1} \quad \text{if } -1 \leqslant r \leqslant n-1 \tag{6.81}$$

and hence that

$$\dim S(V) = (n-1)\,2^n + 1 . \qquad (6.82)$$

Proposition 3

Let V be an n - dimensional vector space.

a) The Z - graded Lie superalgebra S(V) is transitive.

b) If $n \geqslant 3$, then

$$\langle S_r(V) , S_1(V) \rangle = S_{r+1}(V) \quad \text{for all } r \geqslant -1 . \qquad (6.83)$$

c) The Lie superalgebra S(V) is simple provided that $n \geqslant 3$.

Proof

Part a) follows from the analogous statement for W(V), to prove b) one may use the subsequent lemma, finally, c) follows from lemma 4 of §2.

Lemma 4

Let V be a finite-dimensional vector space and let $r \geqslant 0$ be an integer. The union of the following two sets of superderivations of $\wedge V$ generates the vector space $S_r(V)$.

(I)

$$x_1 \wedge \ldots\ldots \wedge x_{r+1} \wedge D_g \qquad (6.84,a)$$

with $x_1, \ldots\ldots, x_{r+1} \in V$; $g \in V^*$

$$g(x_s) = 0 \quad \text{for } 1 \leqslant s \leqslant r+1 \qquad (6.84,b)$$

(II)

$$x \wedge x_1 \wedge \ldots\ldots \wedge x_r \wedge D_f - y \wedge x_1 \wedge \ldots\ldots \wedge x_r \wedge D_g \qquad (6.85,a)$$

with $x, y, x_1, \ldots\ldots, x_r \in V$; $f, g \in V^*$

$$f(x) = g(y) = 1 \;, \quad f(y) = g(x) = 0 \qquad (6.85,b)$$

$$f(x_s) = g(x_s) = 0 \quad \text{for } 1 \leqslant s \leqslant r . \qquad (6.85,c)$$

The proof is straightforward and will be omitted.

Remark 3)

Let A be an isomorphism of the vector space V onto a second vector space V'. We have seen (in remark 1)) that A induces an isomorphism $W(V) \longrightarrow W(V')$ of Z-graded Lie superalgebras. Obviously, this isomorphism maps $S(V)$ onto $S(V')$. In particular, the Z-graded Lie superalgebra $S(V)$ is isomorphic to $S(K^n)$; in the following the latter algebra will be denoted by $S(n)$.

Proposition 4

Let $G = \bigoplus_{r \geq -1} G_r$ be a transitive consistently Z-graded Lie superalgebra such that, for some positive integer n,

$$\dim G_{-1} = n \quad , \quad G_0 \simeq sl(n) \quad , \quad G_1 \neq \{0\} . \tag{6.86}$$

Then we have $n \geq 3$ and the Z-graded Lie superalgebras G and $S(n)$ are isomorphic.

Proof

Set $G^*_{-1} = V$. According to theorem 1 there exists an injective homomorphism

$$\eta : G \longrightarrow W(V) \tag{6.87}$$

of Z-graded Lie superalgebras such that

$$\eta(G_{-1}) = W_{-1}(V) = S_{-1}(V) . \tag{6.88}$$

In the case $n = 1$ we have $G_0 = \{0\}$. Because of the transitivity of G this implies that $G_1 = \{0\}$, contrary to our assumption.

Hence we may assume that $n \geq 2$. Then $G_0 \simeq sl(n)$ is a simple Lie algebra. In view of the dimensions this implies that

$$\eta(G_0) = S_0(V) . \tag{6.89}$$

Let $1 \leq r \leq n-1$. The equation (6.89) shows that $\eta(G_r)$ is an $S_0(V)$-submodule of

$$W_r(V) = S_r(V) \oplus T_r(V) . \tag{6.90}$$

From the discussion in $n^o 1$, B we know that the $S_0(V)$-modules $S_r(V)$ and $T_r(V)$ are irreducible (except for $r = n-1$ in which case $S_{n-1}(V) = \{0\}$)

and non-isomorphic. Thus $\eta(G_r)$ is equal to one of the four vector spaces $\{0\}$, $S_r(V)$, $T_r(V)$, $W_r(V)$.

Since $T_0(V)$ is not contained in $\eta(G_0) = S_0(V)$ we conclude from lemma 2 that $T_r(V)$ cannot be contained in $\eta(G_r)$.

Thus we have shown that

$$\eta(G) \subset S(V) . \tag{6.91}$$

According to our assumption we have $\eta(G_1) \neq \{0\}$. This implies that $n \geq 3$ and that

$$\eta(G_1) = S_1(V) . \tag{6.92}$$

But then it follows from proposition 3 that

$$\eta(G_r) = S_r(V) \quad \text{for all } r \geq -1 . \tag{6.93}$$

This proves our proposition.

B. Elementary properties of $\tilde{S}(V,t)$, $\dim V$ even

We have seen that the sector of degree $n-1$ of the algebra $S(V)$ is equal to $\{0\}$. This fact is somewhat remarkable since the representations of $S_0(V)$ in $W_{-1}(V)$ and in $W_{n-1}(V)$ are equivalent.

Let us suppose in the present subsection that

$$n = \dim V \text{ is even} , \; n \geq 2 . \tag{6.94}$$

Then it is possible to "combine the two sectors $W_{-1}(V)$ and $W_{n-1}(V)$", as follows. Choose any non-zero element

$$t \in \overset{n}{\Lambda} V . \tag{6.95}$$

It is easy to check that the linear combinations of the superderivations of the types

$$(1+t) \wedge D_g \quad , \quad (D_g a) \wedge D_g \tag{6.96}$$

$$\text{with } g \in V^* ; \; a \in \overset{r}{\Lambda} V , \; 2 \leq r \leq n$$

form a Z_2 - graded subalgebra of $W(V)$; this Lie superalgebra will be called $\tilde{S}(V,t)$.

We define

$$\tilde{S}_{-1}(V,t) = \{(1+t)\wedge D_g \mid g \in V^*\} \qquad (6.97,a)$$

$$\tilde{S}_r(V,t) = S_r(V) \text{ if } r \in Z ,\ r \neq -1 . \qquad (6.97,b)$$

Evidently,

$$\tilde{S}(V,t) = \bigoplus_{r=-1}^{n-2} \tilde{S}_r(V,t) . \qquad (6.98)$$

It follows that

$$\dim \tilde{S}_{\bar{0}}(V,t) = (n-1)\, 2^{n-1} + 1 \qquad (6.99)$$

$$\dim \tilde{S}_{\bar{1}}(V,t) = (n-1)\, 2^{n-1} . \qquad (6.100)$$

By definition, the Lie algebra $S_0(V) \simeq sl(V)$ is contained in $\tilde{S}_{\bar{0}}(V,t)$. Obviously, for every $r \in Z$, the $S_0(V)$ - modules $S_r(V)$ and $\tilde{S}_r(V,t)$ are isomorphic. Let us stress, however, that $(\tilde{S}_r(V,t))_{r\in Z}$ is a Z - gradation only of the vector space but *not* of the algebra $\tilde{S}(V,t)$.

Proposition 5

Suppose that $n = \dim V$ is even, $n \geq 2$, and that t is a non-zero element of $\overset{n}{\wedge} V$. Then the Lie superalgebra $\tilde{S}(V,t)$ is simple.

The proof is straightforward and may be left to the reader.

Remark 4)

Let A be an isomorphism of the vector space V onto a second vector space V' and let $\hat{A} : \wedge V \longrightarrow \wedge V'$ be the isomorphism of Z - graded algebras which extends A . The corresponding isomorphism $W(V) \longrightarrow W(V')$ of Z - graded Lie superalgebras (see remark 1), equation (6.21)) maps $\tilde{S}(V,t)$ onto $\tilde{S}(V',\hat{A}(t))$. In particular, the Lie superalgebra $\tilde{S}(V,t)$ is isomorphic to $\tilde{S}(K^n, e_1 \wedge \ldots\ldots \wedge e_n)$, $(e_i)_{1\leq i\leq n}$ being the canonical basis of K^n ; in the following, the latter algebra will be denoted by $\tilde{S}(n)$.

C. Filtered Lie superalgebras whose associated Z - graded Lie superalgebra is isomorphic to $S(n)$

Recall that with any filtered Lie superalgebra L there is associated a Z - graded Lie superalgebra $\mathrm{gr}\,L$ (see §1, n°4).

Proposition 6

Let L be a filtered Lie superalgebra. Suppose that, for some positive integer n, the Z - graded Lie superalgebras $\mathrm{gr}\,L$ and $S(n)$ are isomorphic. If n is odd then the Lie superalgebra L is isomorphic to $S(n)$, if n is even then the Lie superalgebra L is isomorphic to $S(n)$ or to $\tilde{S}(n)$.

Proof

Let $(L^r)_{r\in Z}$ be the filtration of L. The Z - gradation of $S(n)$ is consistent with the Z_2 - gradation. Consequently, we have

$$L^r_{\bar{0}} = L^{r+1}_{\bar{0}} \quad \text{if } r \text{ is odd} \tag{6.101,a}$$

$$L^r_{\bar{1}} = L^{r+1}_{\bar{1}} \quad \text{if } r \text{ is even} \tag{6.101,b}$$

(see equation (1.72)).

The cases $n = 0$ and $n = 1$ are trivial, hence we may assume that $n \geqslant 2$. By assumption, the Lie algebra

$$\mathrm{gr}_0 L \simeq L^0_{\bar{0}} / L^1_{\bar{0}} \tag{6.102}$$

is isomorphic to $sl(n)$. Evidently, $L^1_{\bar{0}}$ is a nilpotent ideal of the Lie algebra $L^0_{\bar{0}} = L_{\bar{0}}$. Thus $L^1_{\bar{0}}$ is the radical of $L^0_{\bar{0}}$. According to the Levi, Malcev theorem this implies that there exists a subalgebra G_0 of $L^0_{\bar{0}}$ such that

$$L^0_{\bar{0}} = G_0 \oplus L^1_{\bar{0}} \ . \tag{6.103}$$

Of course, G_0 is isomorphic to $sl(n)$ and $L^0_{\bar{0}}$ is a semi-direct product of the subalgebra G_0 with the ideal $L^1_{\bar{0}}$.

Consider L as a G_0 - module. We know that the subspaces L^r_ξ, with $\xi \in Z_2$ and $r \in Z$, are G_0 - invariant. Since G_0 is semi-simple we can choose a family $(G_r)_{r\in Z}$ of G_0 - invariant subspaces of L such that

$$L_{\bar{0}}^{r} = G_r \oplus L_{\bar{0}}^{r+1} \quad \text{if } r \text{ is even} \tag{6.104,a}$$

$$L_{\bar{1}}^{r} = G_r \oplus L_{\bar{1}}^{r+1} \quad \text{if } r \text{ is odd .} \tag{6.104,b}$$

(Of course, we define the subspace G_0 to be equal to the subalgebra G_0 which has been chosen above.)

Obviously, we have

$$L = \bigoplus_{r \in Z} G_r \ . \tag{6.105}$$

Let

$$\alpha_r : G_r \longrightarrow L^r / L^{r+1} \tag{6.106}$$

be the bijective linear mapping which is induced by the canonical mapping $L^r \longrightarrow L^r/L^{r+1}$ and let

$$\alpha : L \longrightarrow \operatorname{gr} L \tag{6.107}$$

be the bijective linear mapping which is defined by the family $(\alpha_r)_{r \in Z}$.

By assumption, there exists an isomorphism

$$\beta : \operatorname{gr} L \longrightarrow S(n) \tag{6.108}$$

of Z - graded Lie superalgebras.

On the other hand we may apply theorem 2 and proposition 2 of section 1 : There exists an injective homomorphism

$$\theta : L \longrightarrow W(n) \tag{6.109}$$

of Lie superalgebras which defines an isomorphism of the filtered Lie superalgebra L onto a filtered subalgebra of $W(n)$ and which maps G_0 into $S_0(n)$. It follows that

$$\theta(G_0) = S_0(n) \ ; \tag{6.110}$$

consequently, $\theta(L)$ is an $S_0(n)$ - submodule of $W(n)$.

Set

$$\delta = \theta \circ \alpha^{-1} \circ \beta^{-1} \tag{6.111}$$

and let δ_0 be the automorphism of the Lie algebra $S_0(n)$ which is induced by the mapping δ. Then δ is an injective linear mapping which satisfies

$$\delta(\langle a, y\rangle) = \langle \delta_0(a), \delta(y)\rangle \quad \text{for all } a \in S_0(n),\ y \in S(n). \tag{6.112}$$

But then our results on the $S_0(n)$-module $W(n)$ imply that

$$\theta(G_r) = \delta(S_r(n)) = S_r(n) \quad \text{if } 0 \leq r \leq n-2 \tag{6.113,a}$$

$$\theta(G_{-1}) = \delta(S_{-1}(n)) \subset S_{-1}(n) \oplus T_{n-1}(n) \tag{6.113,b}$$

(where, of course, $T_{n-1}(n) = T_{n-1}(K^n)$).

Let us first assume that n is odd. Since $\theta(G_{-1})$ is an odd subspace of $W(n)$ the relation (6.113,b) implies that

$$\theta(G_{-1}) \subset S_{-1}(n). \tag{6.114}$$

It follows that

$$\theta(L) = S(n), \tag{6.115}$$

as required.

Next we shall consider the case where n is even. We define two linear mappings

$$\lambda : G_{-1} \longrightarrow S_{-1}(n) \quad ; \quad \lambda' : G_{-1} \longrightarrow T_{n-1}(n) \tag{6.116,a}$$

by the condition that

$$\theta(X) = \lambda(X) + \lambda'(X) \quad \text{for all } X \in G_{-1}. \tag{6.116,b}$$

The mapping λ is bijective, moreover, it is easy to see that $\lambda' \circ \lambda^{-1}$ is $S_0(n)$-invariant. Consequently, there exists an element $t \in \overset{n}{\Lambda}(K^n)$ such that

$$\theta(X) = (1+t) \wedge \lambda(X) \quad \text{for all } X \in G_{-1}. \tag{6.117}$$

If $t = 0$ we conclude that

$$\theta(L) = S(n) \tag{6.118}$$

whereas for $t \neq 0$ we have

$$\theta(L) = \tilde{S}(K^n, t) \simeq \tilde{S}(n). \tag{6.119}$$

3. The Lie superalgebras $\tilde{H}(\psi)$ and $H(\psi)$

As before, V denotes an n-dimensional vector space.

A. Elementary properties of $\tilde{H}(\psi)$ and $H(\psi)$

The equation

$$S(V) = \operatorname{image}(\omega \circ \sigma) \tag{6.31}$$

leads in a natural way to the definition of certain subalgebras of $S(V)$ which we are going to consider in the present section.

To begin with let us recall that the vector space $\mathrm{Sym}_2(V^*)$ is canonically isomorphic to the space $B^s(V)$ of symmetric bilinear forms on V. In fact, if $f, g \in V^*$, the canonical image of the element $f \cdot g \in \mathrm{Sym}_2(V^*)$ is the bilinear form

$$(x,y) \longrightarrow f(x)\,g(y) + g(x)\,f(y) \quad ; \quad x, y \in V \tag{6.120}$$

on V. Thus we may identify $\mathrm{Sym}_2(V^*)$ with $B^s(V)$.

Let ψ be an arbitrary symmetric bilinear form on V. As an abbreviation we set

$$\omega \circ \sigma(a \otimes \psi) = a * \psi \quad \text{for all } a \in \Lambda V \tag{6.121}$$

(see the equations (6.17) and (6.26) for the definitions of ω and σ, respectively).

This definition can be made more explicit, as follows. In view of the canonical isomorphism $\mathrm{Sym}_2(V^*) \longrightarrow B^s(V)$ there exist a finite index set I as well as linear forms $f_i, g_i \in V^*$; $i \in I$, such that

$$\psi(x,y) = \sum_{i \in I} (f_i(x)\,g_i(y) + g_i(x)\,f_i(y)) \quad \text{for all } x, y \in V . \tag{6.122}$$

Using this expression we obtain

$$a * \psi = \sum_{i \in I} (D_{f_i}(a) \wedge D_{g_i} + D_{g_i}(a) \wedge D_{f_i}) \quad \text{for all } a \in \Lambda V . \tag{6.123}$$

It is now easy to see that

$$\tilde{H}(\psi) = \{ a * \psi \mid a \in \Lambda V \} \tag{6.124}$$

is a Z - graded subalgebra of S(V), where, of course, for all $r \in Z$,

$$\tilde{H}_r(\psi) = \{ a * \psi \mid a \in \overset{r+2}{\Lambda} V \} . \tag{6.125}$$

In fact, if $r, s \in Z$; $r, s \geqslant 1$, and if

$$a = x_1 \wedge \dots \wedge x_r \quad ; \quad b = y_1 \wedge \dots \wedge y_s \tag{6.126}$$
$$\text{with } x_1, \dots, x_r, y_1, \dots, y_s \in V ,$$

then we have

$$\langle a * \psi , b * \psi \rangle = c * \psi \tag{6.127,a}$$

with

$$c = (-1)^r \sum_{p=1}^{r} \sum_{q=1}^{s} (-1)^{p+q} \psi(x_p, y_q) x_1 \wedge \dots \wedge \hat{x}_p \wedge \dots \wedge x_r \wedge y_1 \wedge \dots \wedge \hat{y}_q \wedge \dots \wedge y_s \tag{6.127,b}$$

(where the hat ^ indicates that the factor under it has to be deleted). Moreover, it is not difficult to verify that

$$c = 0 \quad \text{if } r + s \geqslant n + 2 . \tag{6.128}$$

Evidently, we have

$$\tilde{H}(\psi) = \bigoplus_{r=-1}^{n-2} \tilde{H}_r(\psi) . \tag{6.129}$$

On the other hand the relation (6.128) shows that

$$H(\psi) = \bigoplus_{r=-1}^{n-3} \tilde{H}_r(\psi) \tag{6.130}$$

is a Z - graded ideal of $\tilde{H}(\psi)$. According to this definition we have

$$H_r(\psi) = \begin{cases} \tilde{H}_r(\psi) & \text{if } -1 \leqslant r \leqslant n-3 \\ \{0\} & \text{otherwise} . \end{cases} \tag{6.131}$$

<u>Remark</u> 5)

Let A be an isomorphism of the vector space V onto a second vector space V' and let ψ' be the symmetric bilinear form on V' such that

$$\psi(x,y) = \psi'(A(x),A(y)) \quad \text{for all } x, y \in V . \tag{6.132}$$

We have seen (in remark 1)) that A induces an isomorphism $W(V) \longrightarrow W(V')$ of Z-graded Lie superalgebras. Obviously, this isomorphism maps $\tilde{H}(\psi)$ onto $\tilde{H}(\psi')$ and $H(\psi)$ onto $H(\psi')$.

From now on we shall *assume that the symmetric bilinear form* ψ *is non-degenerate.*

Then the canonical isomorphism $gl(V) \longrightarrow W_0(V)$ induces a canonical isomorphism $o(\psi) \longrightarrow \tilde{H}_0(\psi)$ (where $o(\psi)$ denotes the orthogonal Lie algebra defined by ψ); this isomorphism will be used to identify the Lie algebra $\tilde{H}_0(\psi)$ with $o(\psi)$.

Furthermore, the linear mappings

$$\overset{r+2}{\wedge} V \longrightarrow \tilde{H}_r(\psi) \quad , \quad a \longrightarrow a * \psi \tag{6.133}$$

are $o(\psi)$-invariant and bijective provided that $r \neq -2$. In particular, we conclude that

$$\tilde{H}_{-1}(\psi) = W_{-1}(V) \tag{6.134}$$

and that

$$\dim \tilde{H}_r(\psi) = \binom{n}{r+2} \quad \text{if } -1 \leqslant r \leqslant n-2 \tag{6.135}$$

$$\dim \tilde{H}(\psi) = 2^n - 1 \tag{6.136}$$

$$\dim H(\psi) = 2^n - 2 \quad \text{if } n \geqslant 1 . \tag{6.137}$$

Proposition 7

Let V be an n-dimensional vector space and let ψ be a symmetric non-degenerate bilinear form on V.

a) The Z-graded Lie superalgebras $\tilde{H}(\psi)$ and $H(\psi)$ are transitive.

b) We have

$$\langle \tilde{H}_r(\psi), \tilde{H}_{-1}(\psi) \rangle = \tilde{H}_{r-1}(\psi) \quad \text{for all } r \leqslant n-2 . \tag{6.138}$$

c) If $n \geq 4$, then

$$\langle H_r(\psi), H_1(\psi)\rangle = H_{r+1}(\psi) \quad \text{for all } r \geq -1 . \tag{6.139}$$

d) The algebra $H(\psi)$ is the commutator algebra of $\tilde{H}(\psi)$ (see chapter I, §1, example 1)).

e) The Lie superalgebra $H(\psi)$ is simple provided that $n \geq 4$.

Proof

Part a) follows from the analogous result for $W(V)$, the statements b) - d) are easily derived from equation (6.127), finally, e) follows from lemma 4 of §2.

Remark 6)

Let $n \geq 0$ be a positive integer and let ψ_n be that symmetric bilinear form on K^n for which the canonical basis of K^n is orthonormal. In the following, the algebras $\tilde{H}(\psi_n)$ and $H(\psi_n)$ will be denoted by $\tilde{H}(n)$ and $H(n)$, respectively.

Suppose now that the field K is algebraically closed. If ψ is a symmetric non-degenerate bilinear form on an n - dimensional vector space we infer from remark 5) that the Z - graded Lie superalgebras $\tilde{H}(\psi)$ and $H(\psi)$ are isomorphic to $\tilde{H}(n)$ and $H(n)$, respectively.

B. A characterization of the algebras $\tilde{H}(\psi)$ and $H(\psi)$

We remind the reader that in section 1, C we have discussed the mappings

$$\omega : (\Lambda V^*) \otimes V \longrightarrow W(V^*) \tag{6.140}$$

$$\Omega : (\Lambda V^*) \otimes V \longrightarrow \mathrm{Hom}(\Lambda V, V) \tag{6.141}$$

which are (by definition) isomorphisms of Z - graded Lie superalgebras.

Now let $\tilde{\psi}$ be a symmetric non-degenerate bilinear form on V^* and let ψ be the symmetric non-degenerate bilinear form on V which is determined by $\tilde{\psi}$. We recall the definition of ψ:

For every element $x \in V$ there exists a unique element $\tilde{x} \in V^*$ such that

$g(x) = \tilde{\psi}(\tilde{x},g)$ for all $g \in V^*$. The bilinear form ψ is then defined by $\psi(x,y) = \tilde{\psi}(\tilde{x},\tilde{y})$ for all $x, y \in V$.

It is not hard to determine the subalgebra $\Omega \circ \omega^{-1}(\tilde{H}(\tilde{\psi}))$ of $\mathrm{Hom}(\Lambda V, V)$. The result may be described as follows.

Let r be an integer. By definition the sector of degree r of $\mathrm{Hom}(\Lambda V, V)$ may be identified with the vector space $\mathrm{Hom}(\overset{r+1}{\Lambda} V, V)$. If $r \geq -1$, an element $A \in \mathrm{Hom}(\overset{r+1}{\Lambda} V, V)$ belongs to $\Omega \circ \omega^{-1}(\tilde{H}_r(\tilde{\psi}))$ if and only if

$$(x_1, \ldots\ldots, x_{r+2}) \longrightarrow \psi(A(x_1 \wedge \ldots\ldots \wedge x_{r+1}), x_{r+2}) \qquad (6.142)$$
$$\text{with } x_1, \ldots\ldots, x_{r+2} \in V$$

is an alternating $r+2$-linear form on V (for $r = -1$ we set $x_1 \wedge \ldots \wedge x_{r+1}$ equal to 1).

We are now ready to prove the following proposition.

<u>Proposition</u> 8

Let $G = \bigoplus_{r \geq -1} G_r$ be a finite-dimensional transitive consistently Z-graded Lie superalgebra with $G_1 \neq \{0\}$. Suppose that there exists a symmetric non-degenerate bilinear form ψ on G_{-1} such that the (faithful) representation of G_0 in G_{-1} maps the Lie algebra G_0 *onto* the orthogonal Lie algebra $o(\psi)$. Let $\tilde{\psi}$ be the symmetric non-degenerate bilinear form on G_{-1}^* which is determined by ψ.

Then we have $\dim G_{-1} \geq 3$ and either the Z-graded Lie superalgebra G is isomorphic to $\tilde{H}(\tilde{\psi})$ or to $H(\tilde{\psi})$ or else G is of the type b'(4) (see §4, n°3, B).

We say that G is of the type b'(4) if, for any algebraically closed extension field E of K, the Z-graded Lie superalgebra $E \underset{K}{\otimes} G$ over E is isomorphic to the Z-graded Lie superalgebra b'(4) constructed over E.

Proof

We set $\dim G_{-1} = n$. Let

$$\tilde{\mu} : G \longrightarrow \mathrm{Hom}(\Lambda G_{-1}, G_{-1}) \qquad (6.143)$$

be the injective homomorphism of Z - graded Lie superalgebras which has been constructed in section 1, C . Our preparatory result implies that

$$\tilde{\mu}(G) \subset \Omega \circ \omega^{-1}(\tilde{H}(\tilde{\psi})) \quad . \tag{6.144}$$

Consequently, the mapping

$$\eta = \omega \circ \Omega^{-1} \circ \tilde{\mu} \tag{6.145}$$

is an injective homomorphism of Z - graded Lie superalgebras which maps G into $\tilde{H}(\tilde{\psi})$.

Obviously, we have

$$\eta(G_{-1}) = \tilde{H}_{-1}(\tilde{\psi}) \tag{6.146}$$

$$\eta(G_0) = \tilde{H}_0(\tilde{\psi}) \quad . \tag{6.147}$$

The relation $G_1 \neq \{0\}$ implies that $\tilde{H}_1(\tilde{\psi}) \neq \{0\}$ and hence that $n \geqslant 3$; moreover, in the case $n = 3$ we conclude that $\eta(G) = \tilde{H}(\tilde{\psi})$.

Thus we now may assume that $n \geqslant 4$. On the usual identifications the representation of $\tilde{H}_0(\tilde{\psi})$ in $\tilde{H}_1(\tilde{\psi})$ is nothing but the canonical representation of $o(\tilde{\psi})$ in $\overset{3}{\wedge} G^*_{-1}$. It is well-known that this representation is irreducible for all $n \geqslant 3$ except, possibly, for $n = 6$.

Remark 7)

Let W be any vector space whose dimension is even, $\dim W = 2m$, and let ϕ be a symmetric non-degenerate bilinear form on W . Let $(e_j)_{1 \leqslant j \leqslant 2m}$ be any basis of W and let $\delta = \det(\phi(e_i, e_j))$ be the discriminant of ϕ with respect to this basis. It is well-known that the canonical representation of $o(\phi)$ in $\overset{m}{\wedge} W$ is reducible if and only if the square roots of $(-1)^m \delta$ belong to K .

Suppose that this condition is fulfilled. Then $\overset{m}{\wedge} W$ decomposes into the direct sum of two irreducible non-isomorphic $o(\phi)$ - submodules.

After this interruption we proceed with the proof of proposition 8 . If the representation of $\tilde{H}_0(\tilde{\psi})$ in $\tilde{H}_1(\tilde{\psi})$ is irreducible the presupposed relation $G_1 \neq \{0\}$ implies that

$$\eta(G_1) = \tilde{H}_1(\tilde{\psi}) = H_1(\tilde{\psi}) . \tag{6.148}$$

More generally, let us assume that this equation is fulfilled. Then it is easy to derive from proposition 7,c) that

$$\eta(G_r) = H_r(\tilde{\psi}) \quad \text{if } -1 \leq r \leq n-3 . \tag{6.149}$$

On the other hand, the relation $\dim \tilde{H}_{n-2}(\tilde{\psi}) = 1$ implies that

$$\eta(G_{n-2}) = \{0\} \quad \text{or} \quad \eta(G_{n-2}) = \tilde{H}_{n-2}(\tilde{\psi}) . \tag{6.150}$$

In the first case we have $\eta(G) = H(\tilde{\psi})$, in the second case it follows that $\eta(G) = \tilde{H}(\tilde{\psi})$.

It remains to discuss the case where $\eta(G_1) \neq \tilde{H}_1(\tilde{\psi})$. Then $n = 6$ and $\tilde{H}_1(\tilde{\psi})$ is the direct sum of two irreducible $\tilde{H}_0(\tilde{\psi})$ - submodules $\tilde{H}_1^1(\tilde{\psi})$ and $\tilde{H}_1^3(\tilde{\psi})$,

$$\tilde{H}_1(\tilde{\psi}) = \tilde{H}_1^1(\tilde{\psi}) \oplus \tilde{H}_1^3(\tilde{\psi}) . \tag{6.151}$$

It follows that

$$\langle \tilde{H}_1^i(\tilde{\psi}), \tilde{H}_1^i(\tilde{\psi}) \rangle = \{0\} \tag{6.152}$$

if $i \in \{1,3\}$.

In fact, suppose that equation (6.152) does not hold for one $i \in \{1,3\}$. Since the representation of $\tilde{H}_0(\tilde{\psi})$ in $\tilde{H}_2(\tilde{\psi})$ is irreducible we conclude that for this i

$$\langle \tilde{H}_1^i(\tilde{\psi}), \tilde{H}_1^i(\tilde{\psi}) \rangle = \tilde{H}_2(\tilde{\psi}) . \tag{6.153}$$

But then proposition 7,b) implies that

$$\tilde{H}_1(\tilde{\psi}) = \langle\langle \tilde{H}_1^i(\tilde{\psi}), \tilde{H}_1^i(\tilde{\psi}) \rangle, \tilde{H}_{-1}(\tilde{\psi}) \rangle \subset \tilde{H}_1^i(\tilde{\psi}) , \tag{6.154}$$

a contradiction.

Consequently, for $i \in \{1,3\}$,

$$\tilde{H}_{-1}(\tilde{\psi}) \oplus \tilde{H}_0(\tilde{\psi}) \oplus \tilde{H}_1^i(\tilde{\psi}) \tag{6.155}$$

is a Z - graded subalgebra of $\tilde{H}(\tilde{\psi})$.

A similar argument shows that

$$G_r = \{0\} \quad \text{if } r \geqslant 2 \ . \tag{6.156}$$

The equation (6.156) is obvious if $r \geqslant n-1 = 5$. Suppose that

$$G_s \neq \{0\} \quad \text{for some } s \ , \ 2 \leqslant s \leqslant 4 \ . \tag{6.157}$$

Since the representation of $\tilde{H}_0(\tilde{\psi})$ in $\tilde{H}_s(\tilde{\psi})$ is irreducible we conclude that

$$\eta(G_s) = \tilde{H}_s(\tilde{\psi}) \ . \tag{6.158}$$

In view of proposition 7,b) this implies that

$$\tilde{H}_{s-1}(\tilde{\psi}) = \langle \eta(G_s) , \eta(G_{-1}) \rangle \subset \eta(G_{s-1}) \tag{6.159}$$

i.e. that

$$\eta(G_{s-1}) = \tilde{H}_{s-1}(\tilde{\psi}) \ . \tag{6.160}$$

Iterating this argument we find

$$\eta(G_1) = \tilde{H}_1(\tilde{\psi}) \ , \tag{6.161}$$

which is contrary to our assumption.

Since, obviously,

$$\eta(G_1) = \tilde{H}_1^j(\tilde{\psi}) \tag{6.162}$$

for some $j \in \{1,3\}$, we thus have shown that

$$\eta(G) = \tilde{H}_{-1}(\tilde{\psi}) \oplus \tilde{H}_0(\tilde{\psi}) \oplus \tilde{H}_1^j(\tilde{\psi}) \ . \tag{6.163}$$

It remains to prove that the algebras (6.155) are of the type b'(4). To see this we assume that K is algebraically closed and then show that the algebras (6.155) are isomorphic to b'(4). Our assertion follows from the results in §5, we repeat the argument.

Let us choose a Cartan subalgebra of $\tilde{H}_0(\tilde{\psi}) \simeq A_3$ as well as a fundamental system of simple roots; moreover, let us construct the corresponding fundamental weights $\lambda_1 , \lambda_2 , \lambda_3$. (We use the notation introduced in the appendix. The fundamental weights are enumerated according to our conventions for A_3 ; note that these conventions are different from those for D_3 .)

The representation of $\tilde{H}_0(\tilde{\psi})$ in $\tilde{H}_{-1}(\tilde{\psi})$ is equivalent to $\rho(\lambda_2)$ and the representation of $\tilde{H}_0(\tilde{\psi})$ in $\tilde{H}_1^i(\tilde{\psi})$ is equivalent to $\rho(2\lambda_i)$; $i \in \{1,3\}$ (provided that the subspaces $\tilde{H}_1^i(\tilde{\psi})$ have been chosen appropriately). Since the representation $\rho(\lambda_2) \otimes \rho(2\lambda_i)$; $i \in \{1,3\}$, contains the adjoint representation of A_3 exactly once it is now easy to see that the algebras (6.155) are indeed isomorphic to $b'(4)$.

C. Filtered Lie superalgebras whose associated Z - graded Lie superalgebra is isomorphic to $\tilde{H}(\psi)$ or $H(\psi)$

As before, V denotes an n - dimensional vector space and ψ is a symmetric non-degenerate bilinear form on V . We remind the reader that with any filtered Lie superalgebra L there is associated a Z - graded Lie superalgebra $\mathrm{gr}\,L$ (see §1, n°4).

Proposition 9

Let L be a filtered Lie superalgebra such that the Z - graded Lie superalgebra $\mathrm{gr}\,L$ is isomorphic to $\tilde{H}(\psi)$ or to $H(\psi)$. Then the Lie superalgebra L is isomorphic to $\mathrm{gr}\,L$.

Proof

Depending on the case under consideration we define $\hat{H}(\psi) = \tilde{H}(\psi)$ or else $\hat{H}(\psi) = H(\psi)$.

Let $(L^r)_{r \in Z}$ be the filtration of L . The Z gradation of $\hat{H}(\psi)$ is consistent with the Z_2 - gradation. Consequently, we have

$$L^r_{\bar{0}} = L^{r+1}_{\bar{0}} \quad \text{if } r \text{ is odd} \qquad (6.164,a)$$

$$L^r_{\bar{1}} = L^{r+1}_{\bar{1}} \quad \text{if } r \text{ is even} \qquad (6.164,b)$$

(see equation (1.72)).

The cases $n = 0$ and $n = 1$ are trivial.

Let us consider the case $n = 2$. The assertion is trivial if $\hat{H}(\psi) = H(\psi)$, hence we may assume that $\hat{H}(\psi) = \tilde{H}(\psi)$. In this case we have

$$L^{-1} = L \quad , \quad L^{0} = L_{\bar{0}} \quad , \quad L^{r} = \{0\} \text{ if } r \geqslant 1 . \tag{6.165}$$

We shall show that

$$\langle L_{\bar{1}} , L_{\bar{1}} \rangle = \{0\} . \tag{6.166}$$

Let D be a non-zero element of the (one-dimensional) Lie algebra $L_{\bar{0}}$. We have

$$\langle X , Y \rangle = \phi(X,Y)\, D \quad \text{for all } X , Y \in L_{\bar{1}} \tag{6.167}$$

where ϕ is a symmetric bilinear form on $L_{\bar{1}}$.

Suppose that there exists an element $X \in L_{\bar{1}}$ such that $\phi(X,X) \neq 0$. Let Y be any element of $L_{\bar{1}}$ which is orthogonal to X with respect to ϕ. Exploiting the Jacobi identity for the two triples (X,X,X) and (X,X,Y) we obtain

$$\langle D , X \rangle = \langle D , Y \rangle = 0 . \tag{6.168}$$

Thus we have shown that

$$\langle D , L_{\bar{1}} \rangle = \{0\} . \tag{6.169}$$

But this is contrary to the fact that the filtration of L is transitive (see §1, lemma 8).

It follows that $\phi = 0$, hence the equation (6.166) is proved. Obviously, this implies that $L \simeq \mathrm{gr}\, L$, as required.

Therefore, we now may assume that $n \geqslant 3$. By assumption, the Lie algebra

$$\mathrm{gr}_0 L \simeq L_{\bar{0}}^{0} / L_{\bar{0}}^{1} \tag{6.170}$$

is isomorphic to $o(\psi)$. Evidently, $L_{\bar{0}}^{1}$ is a nilpotent ideal of the Lie algebra $L_{\bar{0}}^{0} = L_{\bar{0}}$. Thus $L_{\bar{0}}^{1}$ is the radical of $L_{\bar{0}}^{0}$. According to the Levi, Malcev theorem this implies that there exists a subalgebra G_0 of $L_{\bar{0}}^{0}$ such that

$$L_{\bar{0}}^{0} = G_0 \oplus L_{\bar{0}}^{1} . \tag{6.171}$$

Of course, G_0 is isomorphic to $o(\psi)$ and $L_{\bar{0}}^{0}$ is a semi-direct product of the subalgebra G_0 with the ideal $L_{\bar{0}}^{1}$.

Consider L as a G_0-module. We know that the subspaces L_{ξ}^{r}, with $\xi \in Z_2$ and $r \in Z$, are G_0-invariant. Since G_0 is semi-simple we can choose a family $(G_r)_{r\in Z}$ of G_0-invariant subspaces of L such that

$$L_{\bar{0}}^{r} = G_r \oplus L_{\bar{0}}^{r+1} \quad \text{if } r \text{ is even} \tag{6.172,a}$$

$$L_{\bar{1}}^{r} = G_r \oplus L_{\bar{1}}^{r+1} \quad \text{if } r \text{ is odd .} \tag{6.172,b}$$

(Of course, we define the subspace G_0 to be equal to the subalgebra G_0 which has been chosen above.)

Obviously, we have

$$L = \bigoplus_{r\in Z} G_r . \tag{6.173}$$

For every $r \in Z$ let

$$\tilde{\alpha}_r : L^r \longrightarrow L^r/L^{r+1} \tag{6.174}$$

be the canonical mapping. Let

$$\alpha_r : G_r \longrightarrow L^r/L^{r+1} \tag{6.175}$$

be the bijective linear mapping induced by $\tilde{\alpha}_r$ and let

$$\alpha : L \longrightarrow \operatorname{gr} L \tag{6.176}$$

be the bijective linear mapping defined by the family $(\alpha_r)_{r\in Z}$.

By assumption, there exists an isomorphism

$$\beta : \operatorname{gr} L \longrightarrow A(\psi) \tag{6.177}$$

of Z-graded Lie superalgebras. Set

$$\gamma = \beta \circ \alpha \tag{6.178}$$

and let

$$\gamma_r : G_r \longrightarrow \hat{A}_r(\psi) \quad ; \quad r \in Z \tag{6.179}$$

be the bijective linear mapping which is induced by γ. It is easy to see that

$$\gamma(\langle Q , X\rangle) = \langle \gamma_0(Q) , \gamma(X)\rangle \quad \text{for all } Q \in G_0 , X \in L . \tag{6.180}$$

Next we define a bilinear mapping

$$P : \hat{H}_{-1}(\psi) \times \hat{H}_{-1}(\psi) \longrightarrow \hat{H}(\psi) \tag{6.181,a}$$

by

$$P(\gamma_{-1}(X), \gamma_{-1}(Y)) = \gamma(\langle X , Y\rangle) \quad \text{for all } X , Y \in G_{-1} . \tag{6.181,b}$$

The mapping P is symmetric and $\hat{H}_0(\psi)$ - invariant, moreover, the image of P is contained in the even subspace of $\hat{H}(\psi)$.

Identifying $\hat{H}_0(\psi)$ with $o(\psi)$ we know that the $o(\psi)$ - module $\tilde{H}(\psi)$ only contains the canonical representations of $o(\psi)$ in $\overset{r+2}{\bigwedge} V$, $-1 \leqslant r \leqslant n-2$. On the other hand it is easy to decompose the symmetric tensor product of the $o(\psi)$ - module $\hat{H}_{-1}(\psi)$ with itself. It turns out that this symmetric tensor product and the $o(\psi)$ - module $\tilde{H}(\psi)$ contain as a common simple component only the trivial one-dimensional $o(\psi)$ - module.

We conclude that $P = 0$ if $\hat{H}(\psi) = H(\psi)$ or if $\hat{H}(\psi) = \tilde{H}(\psi)$ and n is odd, whereas

$$P(A,B) \in \tilde{H}_{n-2}(\psi) \quad \text{for all } A , B \in \tilde{H}_{-1}(\psi) \tag{6.182}$$

if $\hat{H}(\psi) = \tilde{H}(\psi)$ and n is even.

Let us show that $P = 0$ in the latter case, too. It is known that there exists a symmetric non-degenerate $\tilde{H}_0(\psi)$ - invariant bilinear form ϕ on $\tilde{H}_{-1}(\psi)$; moreover, any $\tilde{H}_0(\psi)$ - invariant bilinear form on $\tilde{H}_{-1}(\psi)$ is proportional to ϕ. Since $\tilde{H}_0(\psi)$ acts trivially on $\tilde{H}_{n-2}(\psi)$ we conclude that

$$P(A,B) = \phi(A,B)\, D \quad \text{for all } A , B \in \tilde{H}_{-1}(\psi) \tag{6.183}$$

with some element $D \in \tilde{H}_{n-2}(\psi)$. We have to prove that $D = 0$.

Let $X \in G_{-1}$. The relation

$$\gamma(\langle X , X\rangle) = P(\gamma_{-1}(X), \gamma_{-1}(X)) \in \tilde{H}_{n-2}(\psi) \tag{6.184}$$

shows that $\langle X , X\rangle \in G_{n-2}$. On the other hand it follows from the Jacobi identity that

$$\langle\langle X , X\rangle, X\rangle = 0 . \tag{6.185}$$

Obviously, this implies that $\langle\langle X , X\rangle, X\rangle \in G_{n-3}$; moreover, an easy calculation now yields

$$\gamma(\langle\langle X , X\rangle, X\rangle) = \phi(\gamma_{-1}(X), \gamma_{-1}(X))\langle D , \gamma_{-1}(X)\rangle . \tag{6.186}$$

Thus we have shown that

$$\phi(A,A)\langle D , A\rangle = 0 \quad \text{for all } A \in \tilde{H}_{-1}(\psi) . \tag{6.187}$$

Since ϕ is symmetric and non-degenerate and since $\tilde{H}(\psi)$ is transitive we conclude that $D = 0$, as required.

The upshot is that in all cases

$$\langle G_{-1} , G_{-1}\rangle = \{0\} . \tag{6.188}$$

We now apply theorem 2 and proposition 2 of section 1 : There exists an injective homomorphism

$$\theta : L \longrightarrow W(V) \tag{6.189}$$

of Lie superalgebras which defines an isomorphism of the filtered Lie superalgebra L onto a filtered subalgebra of W(V) and which satisfies

$$\theta(G_{-1}) = W_{-1}(V) . \tag{6.190}$$

In the present situation we can even prove that

$$\theta(L) \subset S(V) . \tag{6.191}$$

In fact, suppose that there exists an element $X \in L$ such that $\theta(X)$ is not contained in S(V) . Using lemma 2 of section 1, B as well as equation (6.190) it is not difficult to construct an element $Q \in G_0$ such that, with respect to the decomposition

$$W(V) = \bigoplus_{r=-1}^{n-2} S_r(V) \oplus \bigoplus_{s=0}^{n-1} T_s(V) , \tag{6.192}$$

$\theta(Q)$ has a non-zero component in $T_0(V)$. Since the Lie algebra G_0 is semi-simple this is clearly impossible.

Let us now assume that $\hat{H}(\psi) = H(\psi)$. In this case we have $L^{n-3} \neq \{0\}$ but $L^r = \{0\}$ if $r \geq n-2$. It follows that

$$L^{n-3} = G_{n-3} . \tag{6.193}$$

Since θ preserves the Z_2-gradation as well as the filtration we conclude that

$$\theta(L^{n-3}) \subset W_{n-3}(V) \oplus W_{n-1}(V) . \tag{6.194}$$

The relation (6.191) then implies that

$$\theta(L^{n-3}) \subset W_{n-3}(V) \tag{6.195}$$

(it is only at this place that the relation (6.191) is used).

Let us define, for all positive integers $r \geq 0$,

$$\tilde{G}_{n-3-r} = (\mathrm{ad}\, G_{-1})^r L^{n-3} . \tag{6.196}$$

The elements of $\tilde{G}_{n-3-r}$ are homogeneous of degree $(n-3-r) + 2Z$ with respect to the Z_2-gradation. Moreover, it is easy to see that

$$\theta(\tilde{G}_{n-3-r}) \subset W_{n-3-r}(V) \tag{6.197}$$

$$\tilde{G}_{n-3-r} \subset L^{n-3-r} \tag{6.198}$$

for all $r \geq 0$. Using proposition 7,b) we conclude by induction that

$$\beta\, \tilde{\alpha}_{n-3-r}(\tilde{G}_{n-3-r}) = H_{n-3-r}(\psi) \tag{6.199}$$

for all $r \geq 0$, which is to say that

$$L^{n-3-r} = \tilde{G}_{n-3-r} + L^{n-2-r} \tag{6.200}$$

for all $r \geq 0$.

It is now easy to see that the family $(\tilde{G}_r)_{-1 \leq r \leq n-3}$ is a Z-gradation of the Lie superalgebra L which is consistent with the Z_2-gradation and which induces the given filtration $(L^r)_{r \in Z}$ (see §1, n°4). This implies that $L \simeq \mathrm{gr}\, L$, as required.

The case $\hat{H}(\psi) = \tilde{H}(\psi)$ can be treated similarly.

§7 CLASSIFICATION OF A SPECIAL TYPE OF TRANSITIVE Z - GRADED LIE SUPERALGEBRAS [3]

We remind the reader that all Lie superalgebras are assumed to be *finite-dimensional*. Throughout this paragraph we shall suppose that the field K is *algebraically closed*.

Proposition 1

We suppose that the field K is algebraically closed.

Let $G = \bigoplus_{i \geqslant -1} G_i$ be a transitive consistently Z - graded Lie superalgebra satisfying the following conditions :

a) The Lie algebra G_0 is semi-simple.

b) The subspace $G_{-1} \oplus G_0 \oplus G_1$ generates the algebra G .

c) The representations of G_0 in G_{-1} and G_1 are irreducible (in particular, $G_{\pm 1} \neq \{0\}$).

d) The representations of G_0 in G_{-1} and G_1 are *not* contragredient to each other.

e) The representation of G_0 in G_1 is faithful.

Then the Z - graded Lie superalgebra G is isomorphic to one of the following Z - graded Lie superalgebras :

S(n) with $n \geqslant 4$

H(n) with $n \geqslant 5$, $n \neq 6$

b(n) or b'(n) with $n \geqslant 3$.

Remarks

1) We remind the reader that the Z - graded Lie superalgebra b'(n) is obtained from b(n) by an inversion of the Z - gradation (see chapter 0, §2, 5)).

2) The Z - graded Lie superalgebras S(3) and b'(3) are isomorphic.

3) The representation of $H_0(6)$ in $H_1(6)$ is reducible.

4) In connection with our assumption a) we recall proposition 4 and lemma 5 of §1 .

Proof

In view of our assumption e) we conclude from lemma 1 of §1 that the algebra G is bitransitive. Consequently, it is sufficient to prove the following proposition.

Proposition 1'

We suppose that the field K is algebraically closed.

Let $G = \bigoplus_{i=-r}^{s} G_i$; $r , s \geqslant 1$, be a bitransitive consistently Z - graded Lie superalgebra satisfying the conditions a) - d) of proposition 1 .

If $r = 1$ or $s = 1$, then the Z - graded Lie superalgebra G is isomorphic either to one of the Z - graded Lie superalgebras

S(n) with $n \geqslant 4$

H(n) with $n \geqslant 5$, $n \neq 6$

b(n) with $n \geqslant 3$

or else to one of the algebras obtained from the above by an inversion of the Z - gradation.

Proof

We shall use the notation introduced in the appendix. Let h be a Cartan subalgebra of G_0 . The Killing form ϕ of G_0 yields a non-degenerate symmetric bilinear form on h^* which will be denoted by $(\mid)$.

For every linear form γ on h we define the element $H_\gamma \in h$ by the condition that

$$\gamma(H) = \phi(H_\gamma, H) \quad \text{for all } H \in h . \tag{7.1}$$

If γ is a root of G_0 we choose the root vectors $E_{\pm\gamma}$ associated with the roots $\pm\gamma$ such as to satisfy

$$\langle E_{-\gamma}, E_\gamma \rangle = H_\gamma \quad \text{for all roots } \gamma . \tag{7.2}$$

Finally, let $\alpha_1, \ldots\ldots, \alpha_m$ be a fundamental system of simple roots of G_0 and let $\lambda_1, \ldots\ldots, \lambda_m$ be the corresponding system of fundamental weights.

By assumption the algebra G is bitransitive and the representations of G_0 in G_{-1} and G_1 are irreducible. It follows that $G_0 \neq \{0\}$ and that the representations of G_0 in G_{-1} and G_1 are faithful. Let λ (resp. μ) be the highest (resp. lowest) weight of the representation of G_0 in G_{-1} (resp. in G_1) and let $X_\lambda \in G_{-1}$ (resp. $Y_\mu \in G_1$) be a weight vector associated with it. Of course, λ and μ are different from zero.

It is easy to see that

$$\lambda + \mu = -\alpha \tag{7.3}$$

is a root of G_0 and that $\langle X_\lambda , Y_\mu \rangle$ is a (non-zero!) root vector which belongs to $-\alpha$ (see the proof of proposition 4 in §1). Consequently, we may assume that

$$\langle X_\lambda , Y_\mu \rangle = E_{-\alpha} \,. \tag{7.4}$$

Let $G' = \bigoplus_{i=-s}^{r} G'_i$ be the Z - graded Lie superalgebra which is obtained from G by an inversion of the Z - gradation; thus, by definition, we have $G'_i = G_{-i}$ for all $i \in Z$. Evidently, the algebra G' satisfies the assumptions of proposition 1'.

It is well-known that there exists a unique element w_0 of the Weyl group of G_0 such that

$$w_0\{\alpha_1, \ldots\ldots, \alpha_m\} = \{-\alpha_1, \ldots\ldots, -\alpha_m\} \; ; \tag{7.5}$$

moreover, $w_0(\mu)$ (resp. $w_0(\lambda)$) is the highest (resp. lowest) weight of the representation of G'_0 in G'_{-1} (resp. G'_1) and $w_0(\mu) + w_0(\lambda) = -w_0(\alpha)$.

The root $w_0(\alpha)$ is positive (resp. negative) if and only if the root α is negative (resp. positive). Thus, by considering the algebra G' instead of G if necessary, we may assume that the root α is positive. It follows that

$$\langle E_\alpha , X_\lambda \rangle = \langle E_{-\alpha} , Y_\mu \rangle = 0 \;. \tag{7.6}$$

Next we shall prove several lemmas.

Lemma 1

Let σ and τ be linear forms on h and let $S_\sigma \in G_{-1}$ and $T_\tau \in G_1$ be elements such that

$$\langle H, S_\sigma \rangle = \sigma(H) S_\sigma \quad , \quad \langle H, T_\tau \rangle = \tau(H) T_\tau \quad \text{for all } H \in h . \qquad (7.7)$$

Let δ be a root of G_0 such that the following equations are satisfied :

$$\langle S_\sigma, T_\tau \rangle = a E_{-\delta} \quad \text{for some } a \in K \qquad (7.8)$$

$$\langle E_\delta, S_\sigma \rangle = 0 . \qquad (7.9)$$

It then follows that

$$a(\sigma | \delta) = 0 . \qquad (7.10)$$

Proof

Obviously, we may assume that $a \neq 0$. Then the equation (7.8) implies that S_σ and T_τ are non-zero and that

$$\sigma + \tau + \delta = 0 . \qquad (7.11)$$

Let us set

$$T_{-\sigma} = \langle E_\delta, T_\tau \rangle . \qquad (7.12)$$

It is easy to check that

$$\langle S_\sigma, T_{-\sigma} \rangle = -a H_\delta \ ; \qquad (7.13)$$

in particular, we have $T_{-\sigma} \neq 0$.

Consider the two elements

$$\langle S_\sigma, S_\sigma \rangle \in G_{-2} \quad , \quad \langle T_{-\sigma}, T_{-\sigma} \rangle \in G_2 . \qquad (7.14)$$

We have

$$\langle T_{-\sigma}, \langle S_\sigma, S_\sigma \rangle\rangle = -2a(\sigma|\delta) S_\sigma \qquad (7.15)$$

$$\langle S_\sigma, \langle T_{-\sigma}, T_{-\sigma} \rangle\rangle = 2a(\sigma|\delta) T_{-\sigma} . \qquad (7.16)$$

These relations imply the equation (7.10), for otherwise both of the spaces $G_{\pm 2}$ would be different from $\{0\}$.

Corollary 1

We have

$$(\lambda|\alpha) = 0 . \tag{7.17}$$

In particular, G_0 is not isomorphic to the Lie algebra A_1 .

Corollary 2

Let β be a positive root of G_0 .

a) If $\alpha+\beta$ is a root of G_0 , then

$$(\lambda-\beta|\alpha+\beta) = 0 \tag{7.18}$$

$$(\lambda|\beta) > 0 . \tag{7.19}$$

b) If $\alpha+\beta$ is a root of G_0 but $\alpha-\beta$ is not a root, then

$$(\lambda+\alpha|\beta) = 0 \tag{7.20}$$

$$2\frac{(\lambda|\beta)}{(\beta|\beta)} = -2\frac{(\alpha|\beta)}{(\beta|\beta)} = 1 . \tag{7.21}$$

Proof

a) We set

$$S_\sigma = \langle E_{-\beta}, X_\lambda \rangle \quad , \quad T_\tau = Y_\mu . \tag{7.22}$$

Then

$$\langle S_\sigma, T_\tau \rangle = \langle E_{-\beta}, E_{-\alpha} \rangle . \tag{7.23}$$

By assumption, $\alpha+\beta$ is a root of G_0 , hence $\langle E_{-\beta}, E_{-\alpha} \rangle$ is a (non-zero!) root vector of G_0 which belongs to the root $-\delta = -\alpha-\beta$.

On the other hand, it is easy to verify that

$$\langle E_\delta, S_\sigma \rangle = 0 . \tag{7.24}$$

Consequently, lemma 1 yields

$$(\lambda-\beta|\alpha+\beta) = (\sigma|\delta) = 0 . \tag{7.25}$$

Equation (7.23) implies that $\langle E_{-\beta}, X_\lambda \rangle = S_\sigma \neq 0$, whereas, obviously, $\langle E_\beta, X_\lambda \rangle = 0$. This proves the inequality (7.19).

b) We set

$$S_\sigma = \langle E_{-\beta}, X_\lambda \rangle \quad , \quad T_\tau = \langle E_\beta, Y_\mu \rangle \ . \tag{7.26}$$

It is not difficult to see that

$$\langle S_\sigma, T_\tau \rangle = (\lambda + \alpha | \beta)\, E_{-\alpha} \tag{7.27}$$

and that

$$\langle E_\alpha, S_\sigma \rangle = 0 \ . \tag{7.28}$$

Thus we conclude from lemma 1 that

$$(\lambda + \alpha | \beta)(\lambda - \beta | \alpha) = 0 \ . \tag{7.29}$$

According to corollary 1 we have $(\lambda|\alpha) = 0$; on the other hand, our assumptions on β imply that $(\alpha|\beta) < 0$. Thus the equation (7.20) follows from equation (7.29) .

Obviously, the equations (7.17) , (7.18) and (7.20) imply (7.21) .

Lemma 2

Let β and γ be two positive roots of G_0 . Suppose that $\alpha + \beta$ is a root but that $\alpha - \beta$, $\alpha - \gamma$ and $\beta - \gamma$ are all non-zero and not roots. Then

$$(\lambda | \gamma) = 0 \ . \tag{7.30}$$

Proof

We set

$$Y_{-\lambda} = \langle E_\alpha, Y_\mu \rangle \ . \tag{7.31}$$

Then

$$\langle X_\lambda, Y_{-\lambda} \rangle = -H_\alpha \tag{7.32}$$

and hence $Y_{-\lambda} \neq 0$; moreover, we have

$$\langle E_{-\beta}, Y_{-\lambda} \rangle = \langle E_{-\gamma}, Y_{-\lambda} \rangle = 0 \ . \tag{7.33}$$

Let us define

$$A = \langle X_\lambda , \langle E_{-\gamma} , \langle E_{-\beta} , X_\lambda \rangle\rangle\rangle \in G_{-2} \qquad (7.34,a)$$

$$B = \langle Y_{-\lambda} , \langle E_{\gamma} , \langle E_{\beta} , Y_{-\lambda} \rangle\rangle\rangle \in G_{2} \qquad (7.34,b)$$

as well as

$$b = (\alpha|\beta+\gamma)(\lambda|\gamma) - (\alpha|\gamma)(\beta|\gamma) . \qquad (7.35)$$

It is not difficult to verify that

$$\langle E_\beta , \langle E_\gamma , \langle Y_{-\lambda} , A \rangle\rangle\rangle = b(\lambda|\beta) X_\lambda \qquad (7.36,a)$$

$$\langle E_{-\beta} , \langle E_{-\gamma} , \langle X_\lambda , B \rangle\rangle\rangle = -b(\lambda|\beta) Y_{-\lambda} . \qquad (7.36,b)$$

Our assumptions on the roots β and γ imply that

$$(\alpha|\beta) < 0 \quad , \quad (\alpha|\gamma) \leq 0 \quad , \quad (\beta|\gamma) \leq 0 , \qquad (7.37)$$

moreover, it is obvious that $(\lambda|\gamma) \geq 0$.

Suppose that $(\lambda|\gamma) \neq 0$, i.e. that $(\lambda|\gamma) > 0$. Then we deduce from the inequalities (7.37) that $b < 0$. On the other hand, according to the inequality (7.19), we have $(\lambda|\beta) > 0$. Consequently, we derive from the equations (7.36) that A and B are different from zero. This is impossible.

Lemma 3

a) λ is a fundamental weight of G_0 .

b) The Lie algebra G_0 is simple.

c) α is a maximal element of the set of all positive roots δ of G_0 such that $(\lambda|\delta) = 0$.

Proof

a) and b) Let $\hat{G}_0$ be a simple ideal of G_0 . It is well-known that $h \cap \hat{G}_0$ is a Cartan subalgebra of $\hat{G}_0$. The transitivity of G implies that the representation of G_0 in G_{-1} is faithful. Therefore, the restriction of λ to $h \cap \hat{G}_0$ is non-zero.

Suppose now that $\hat{G}_0$ is the simple ideal of G_0 to which the root α belongs. Bearing in mind the equation $(\lambda|\alpha) = 0$ we deduce from the above remark that there exists a simple root β of G_0 such that $\alpha+\beta$ is a root but $\alpha-\beta$ is not. According to the corollary 2 of lemma 1 this implies that

$$2\,\frac{(\lambda|\beta)}{(\beta|\beta)} = 1 . \tag{7.38}$$

Let γ be any simple root of G_0 such that

$$(\lambda|\gamma) \neq 0 . \tag{7.39}$$

Then $\gamma = \beta$.

Suppose the contrary. Obviously, $\alpha-\gamma$ and $\beta-\gamma$ are non-zero and are not roots. Thus lemma 2 shows that $(\lambda|\gamma) = 0$, a contradiction.

In view of equation (7.38) this result proves that λ is a fundamental weight of G_0. But the representation of G_0 in G_{-1} is faithful, hence the Lie algebra G_0 must be simple.

c) Let γ and δ be two roots of G_0. Recall that δ is said to be greater than γ (with respect to the fundamental system $\alpha_1,\ldots\ldots,\alpha_m$) if and only if

$$\delta - \gamma = \sum_{i=1}^{m} c_i\,\alpha_i \quad \text{with } c_i \in Z,\ c_i \geq 0 . \tag{7.40}$$

If this is the case we write $\delta \geq \gamma$.

Suppose now that there exists a positive root δ of G_0 which satisfies

$$(\lambda|\delta) = 0 \tag{7.41}$$

and which is strictly greater than α.

The latter property implies that there exists a positive root γ of G_0 such that $\alpha+\gamma$ is a root of G_0 and that $\delta \geq \alpha+\gamma$. It follows that

$$0 = (\lambda|\alpha) \leq (\lambda|\alpha+\gamma) \leq (\lambda|\delta) = 0 \tag{7.42}$$

and hence that

$$(\lambda|\gamma) = 0 . \tag{7.43}$$

This is contrary to the inequality (7.19).

Lemma 4

There exist only the following two possibilities : Either

a) the fundamental weight λ belongs to an extremal vertex of the Dynkin diagram of G_0

or else

b) α is a simple root of G_0 which belongs to an extremal vertex of the Dynkin diagram and λ belongs to the neighbouring vertex, the latter vertex being non-extremal.

Proof

Let β be the unique simple root of G_0 such that $(\lambda|\beta) \neq 0$. Both λ and β belong to the same vertex v of the Dynkin diagram.

Suppose that v is not extremal. Choose two different vertices v' and v" of the Dynkin diagram which are direct neighbours of v and let β' and β'' be the simple roots of G_0 which belong to v' and v", respectively. Then

$$\gamma = \beta' + \beta + \beta'' \tag{7.44}$$

is a positive root of G_0.

Evidently, $\beta - \gamma$ is different from zero but not a root. Since $(\lambda|\gamma) \neq 0$ but $(\lambda|\alpha) = 0$ we conclude from lemma 2 that $\alpha - \gamma$ is a root of G_0 which, obviously, must be negative. It follows that

$$\alpha = \beta' \quad \text{or} \quad \alpha = \beta'' \; ; \tag{7.45}$$

consequently, α is a simple root of G_0. But then we deduce from lemma 3,c) that α belongs to an extremal vertex of the Dynkin diagram and that λ belongs to the neighbouring vertex.

The next lemma is the crucial intermediate result.

Lemma 5

Let us enumerate the vertices of the Dynkin diagram of the simple Lie algebra G_0 as described in the appendix. Then there exist only the following possibilities :

Case a) of lemma 4

G_0	λ	$w_0(\mu)$
A_n, $n \geq 2$	λ_1	$\lambda_1 + \lambda_{n-1}$
	λ_n	$\lambda_2 + \lambda_n$
B_2	λ_1	$2\lambda_2$
B_3	λ_1	$2\lambda_3$
B_n, $n \geq 4$	λ_1	λ_3
D_4	λ_1	$\lambda_3 + \lambda_4$
	λ_3	$\lambda_1 + \lambda_4$
	λ_4	$\lambda_1 + \lambda_3$
D_n, $n \geq 5$	λ_1	λ_3

Case b) of lemma 4

G_0	λ	$w_0(\mu)$
A_n, $n \geq 3$	λ_2	$2\lambda_n$
	λ_{n-1}	$2\lambda_1$
B_3	λ_2	$2\lambda_3$
C_n, $n \geq 3$	λ_2	$2\lambda_1$

The tables contain the highest weights λ and $w_0(\mu)$ of the representations of G_0 in G_{-1} and G_1, respectively.

Proof

Let β be the unique simple root of G_0 such that $(\lambda|\beta) \neq 0$ and let s be the number of the vertex of the Dynkin diagram to which both λ and β belong; thus

$$\lambda = \lambda_s \quad , \quad \beta = \alpha_s \ . \tag{7.46}$$

According to the corollary 2 of lemma 1 (see also the proof of lemma 3) we have

$$2\,\frac{(\alpha|\beta)}{(\beta|\beta)} = -1 \ . \tag{7.47}$$

We shall discuss the two cases mentioned in lemma 4 . In doing so we have to consider all simple Lie algebras separately. We describe the general procedure and leave the details to the reader.

Case a)

In this case the vertex s of the Dynkin diagram is extremal.

Let (G_0, λ_s) be a pair consisting of a simple Lie algebra $G_0 \neq A_1$ and a fundamental weight λ_s , the vertex s being extremal. The criterion in lemma 3,c) (with $\lambda = \lambda_s$) uniquely determines a positive root α of G_0 ; the root $\beta = \alpha_s$ has already been introduced. If the case $\lambda = \lambda_s$ and $w_0(\mu) = -w_0(\alpha + \lambda_s)$ is not included in the table then in this case one of our necessary conditions is violated.

To see this we first check whether the equation (7.47) is satisfied; we are finished if this is not the case. Otherwise we look for a positive root γ of G_0 such that $\alpha - \gamma$ and $\beta - \gamma$ are non-zero and are not roots but such that $(\lambda|\gamma) \neq 0$. The existence of such a root γ is contrary to our lemma 2 .

Usually, there will be many roots γ which meet these requirements; a possible choice for γ is the following. Let t be an extremal vertex of the Dynkin diagram which is different from s (there may be at most two such vertices). Let γ_t be maximal in the set of all positive roots δ such that $(\lambda_t|\delta) = 0$. Apart from two cases (namely $G_0 = E_8$, $\lambda = \lambda_7$ and $G_0 = G_2$, $\lambda = \lambda_2$) the root $\gamma = \gamma_t$ satisfies our requirements for at least one t .

In the exceptional case for E_8 one may choose $\gamma = \gamma_1 - \alpha_7$. In the exceptional case for G_2 it turns out that $\beta' = \alpha + \beta$ and $\alpha + \beta'$ are positive roots, moreover, we have $(\lambda - \beta'|\alpha + \beta') \neq 0$. This is contrary to the corollary 2,a) of lemma 1 .

Case b)

In this case we have $\alpha = \alpha_t$ where t is an extremal vertex of the Dynkin diagram, moreover, the vertex s is non-extremal and is the direct neighbour of the vertex t .

Let $(G_0, \alpha_t, \lambda_s)$ be a triple consisting of a simple Lie algebra G_0 , a simple root α_t and of a fundamental weight λ_s as described above. If the case $\lambda = \lambda_s$ and $w_0(\mu) = -w_0(\alpha_t + \lambda_s)$ is not included in the table then the necessary condition in lemma 2 is violated.

To see this we have to find a suitable positive root γ (the root $\beta = \alpha_s$ has already been chosen). This can be achieved as follows. Let θ be the highest root of the Lie algebra G_0. If

$$(\theta|\alpha_s) = (\theta|\alpha_t) = 0 \tag{7.48}$$

then we may choose

$$\gamma = \theta . \tag{7.49}$$

On the other hand, if one of the equations (7.48) does not hold we can choose γ to be maximal in the set of all positive roots δ such that $(\lambda_t|\delta) = 0$.

Remarks

5) For $G_0 = A_n$ the possible highest weights $\lambda, w_0(\mu)$ come by pairs whose members are connected by an automorphism of the Dynkin diagram; consequently, the corresponding irreducible representations can be transformed into each other by an automorphism of the Lie algebra. An analogous remark applies to the three possibilities with $G_0 = D_4$.

6) The reader will notice that the two final possibilities contained in the second table do not correspond to algebras mentioned in the proposition 1'. Our necessary conditions are not strong enough to exclude these cases; they will be eliminated in the subsequent discussion.

We now are ready to complete the proof of proposition 1'. The representations of G_0 in G_{-1} and G_1 are equivalent to $\rho(\lambda)$ and $\rho(w_0(\mu))$, respectively, where $(G_0, \lambda, w_0(\mu))$ is one of the triples listed in lemma 5. For any element $Q \in G_0$ let $\tilde{Q}$ (resp. $\hat{Q}$) denote the corresponding homothety of the G_0-module G_{-1} (resp. G_1).

It can be shown that the tensor product $\rho(\lambda) \otimes \rho(w_0(\mu))$ contains the adjoint representation of G_0 exactly once. This means that there exists a non-zero G_0-invariant bilinear mapping

$$P : G_{-1} \times G_1 \longrightarrow G_0 \tag{7.50}$$

and that any such mapping is proportional to P .

In particular, there exists a non-zero constant $a \in K$ such that

$$\langle X , Y \rangle = a\, P(X,Y) \quad \text{for all } X \in G_{-1} , Y \in G_1 . \tag{7.51}$$

It follows that

$$\langle Y ,\langle X , X' \rangle\rangle = a\, \tilde{P}(X,Y)\, X' + a\, \tilde{P}(X',Y)\, X \tag{7.52}$$
$$\text{for all } X , X' \in G_{-1} \text{ and } Y \in G_1 .$$

Consequently, a necessary (and sufficient) condition for G_{-2} to be equal to $\{0\}$ is that

$$\tilde{P}(X,Y)\, X' + \tilde{P}(X',Y)\, X = 0 \quad \text{for all } X , X' \in G_{-1} \text{ and } Y \in G_1 . \tag{7.53}$$

Similarly, the subspace G_2 vanishes (if and) only if

$$\hat{P}(X,Y)\, Y' + \hat{P}(X,Y')\, Y = 0 \quad \text{for all } X \in G_{-1} \text{ and } Y , Y' \in G_1 . \tag{7.54}$$

Let us stress that the validity of these conditions can be checked once the Lie algebra G_0 and the two highest weights λ and $w_0(\mu)$ (i.e. the G_0 - modules $G_{\pm 1}$) are given.

A straightforward explicit calculation now shows that for the two final cases in the second table of lemma 5 none of the two conditions (7.53) and (7.54) is satisfied. Consequently, these cases have to be excluded.

Thus, according to lemma 5 and to the last result, there exists among the Z - graded Lie superalgebras

$S(n)$ with $n \geq 4$

$H(n)$ with $n \geq 5$, $n \neq 6$

$b'(n)$ with $n \geq 3$

exactly one, denoted by $\bar{G} = \bigoplus_{i \geq -1} \bar{G}_i$, such that the following statement holds :

There exist an isomorphism of Lie algebras

$$g_0 : G_0 \longrightarrow \bar{G}_0 \tag{7.55,a}$$

as well as two bijective linear mappings

$$g_{\pm 1} : G_{\pm 1} \longrightarrow \bar{G}_{\pm 1} \tag{7.55,b}$$

such that

$$g_{\pm 1}(\langle Q , X \rangle) = \langle g_0(Q) , g_{\pm 1}(X) \rangle \quad \text{for all } Q \in G_0 \text{ and } X \in G_{\pm 1} . \tag{7.56}$$

Obviously, this implies that the bilinear mapping

$$\bar{P} : G_{-1} \times G_1 \longrightarrow G_0 \tag{7.57,a}$$

which is defined by

$$\bar{P}(X,Y) = g_0^{-1}(\langle g_{-1}(X) , g_1(Y) \rangle) \quad \text{for all } X \in G_{-1} \text{ and } Y \in G_1 \tag{7.57,b}$$

is G_0 - invariant. According to the discussion above we conclude that there exists a non-zero element $c \in K$ such that

$$g_0(\langle X , Y \rangle) = c \langle g_{-1}(X) , g_1(Y) \rangle \quad \text{for all } X \in G_{-1} \text{ and } Y \in G_1 . \tag{7.58}$$

Of course, we now may even assume that $c = 1$. Bearing in mind the propositions 3 and 7 of §6 we can apply the corollary to proposition 1 in §1 to conclude that the Z - graded Lie superalgebras G and $\bar{G}$ are isomorphic.

§ 8 THE MAIN CLASSIFICATION THEOREMS

We remind the reader that all Lie superalgebras are assumed to be *finite-dimensional*. Throughout this paragraph we shall suppose that the field K is *algebraically closed*.

Now, at last, we are ready to prove the two main classification theorems [3].

Theorem 1 (Kac)

Suppose that the field K is algebraically closed.

Let $G = \bigoplus_{i \geq -1} G_i$ be a transitive irreducible consistently Z-graded Lie superalgebra with $G_1 \neq \{0\}$. Then the Z-graded Lie superalgebra G is isomorphic to one of the following Z-graded Lie superalgebras :

(I) spl(n,m) with $n > m \geq 1$

spl(n,n) / $K \cdot I_{2n}$ with $n \geq 2$

osp(2,2r) with $r \geq 2$

b(n) or b'(n) with $n \geq 3$

(II) W(n) with $n \geq 3$

S(n) with $n \geq 4$

$\tilde{H}(n)$ with $n \geq 4$

H(n) with $n \geq 5$

(III) H^{ξ} with a simple Lie algebra H

(IV) G^{Z} with a Z-graded Lie superalgebra G of the types (I), (II) or (III) for which the center of G_0 is trivial.

Remarks

1) We remind the reader that the Z-graded Lie superalgebra b'(n) is obtained from b(n) by an inversion of the Z-gradation (see chapter 0, §2, 5)).

2) The algebras H^{ξ} and G^{Z} have been defined in §1, n°3.

3) We have the following isomorphisms of Z - graded Lie superalgebras :

$$W(2) \simeq spl(2,1) \simeq osp(2,2) \tag{8.1}$$

$$S(3) \simeq b'(3) \tag{8.2}$$

$$\tilde{H}(3) \simeq o(3)^{\xi} \tag{8.3}$$

$$H(4) \simeq spl(2,2) / K \cdot I_4 \ . \tag{8.4}$$

Proof

All of the Lie superalgebras that we are discussing in this proof are Z - graded. Let us agree (for this proof) to simply call two such algebras isomorphic (without a further specification) if in fact they are isomorphic as Z - graded Lie superalgebras.

According to lemma 5 of §1 we may assume that

$$\langle G_{-1}, G_1 \rangle = G_0 \ ; \tag{8.5}$$

we have to show that on this additional assumption G is isomorphic to one of the algebras (I) - (III).

Let us recall (see §1, proposition 3) that :

α) The Lie algebra G_0 is reductive and the semi-simple part $\langle G_0, G_0 \rangle$ of G_0 is non-zero.

β) The center G_0^0 of G_0 is at most one-dimensional. If $\dim G_0^0 = 1$, then there exists a unique element $C \in G_0^0$ such that

$$\langle C, X \rangle = jX \quad \text{for all } X \in G_j \text{ and all } j \in Z \ . \tag{8.6}$$

γ) The representation of G_0 in G_1 is completely reducible.

In view of γ) the G_0 - module G_1 decomposes into the direct sum of irreducible submodules $G_1^{(s)}$, $1 \leq s \leq r$,

$$G_1 = \bigoplus_{s=1}^{r} G_1^{(s)} \ . \tag{8.7}$$

Evidently,

$$G_{-1} \oplus G_0 \oplus G_1^{(s)} \oplus \langle G_1^{(s)}, G_1^{(s)} \rangle \oplus \dots . \tag{8.8}$$

is a transitive irreducible Z - graded subalgebra of G . Consequently, we infer from proposition 2 of §1 that

$$\langle G_0 , G_0 \rangle \subset \langle G_{-1} , G_1^{(s)} \rangle \quad \text{if } 1 \leqslant s \leqslant r . \tag{8.9}$$

Let us define

$$G^{(s)} = G_{-1} \oplus \langle G_{-1} , G_1^{(s)} \rangle \oplus G_1^{(s)} \oplus \langle G_1^{(s)}, G_1^{(s)} \rangle \oplus \ldots . \tag{8.10}$$

It follows that :

a) $G^{(s)}$ is a transitive Z - graded subalgebra of G .

b) The representations of $G_0^{(s)} = \langle G_{-1} , G_1^{(s)} \rangle$ in G_{-1} and in $G_1^{(s)}$ are irreducible.

c) The subspace $G_{-1} \oplus G_1^{(s)}$ generates the algebra $G^{(s)}$.

Now there exist two possibilities :

1) The representations of $G_0^{(s)}$ in G_{-1} and in $G_1^{(s)}$ are contragredient to each other.

Then $G^{(s)}$ is isomorphic to one of the algebras

spl(n,m) with $n > m \geqslant 1$

spl(n,n) / $K \cdot I_{2n}$ with $n \geqslant 2$

osp(2,2r) with $r \geqslant 2$

(see §5, proposition 2).

2) The representations of $G_0^{(s)}$ in G_{-1} and in $G_1^{(s)}$ are *not* contragredient to each other.

In this case $G_0^{(s)}$ is a simple Lie algebra (see §1, proposition 4). Consequently, if the representation of $G_0^{(s)}$ in $G_1^{(s)}$ is not faithful, then this representation is one-dimensional and trivial. According to lemma 6 of §1 this implies that

$$G^{(s)} \simeq (G_0^{(s)})^{\xi} . \tag{8.11}$$

On the other hand, if the representation of $G_0^{(s)}$ in $G_1^{(s)}$ is faithful, then proposition 1 of §7 shows that $G^{(s)}$ is isomorphic to one of the algebras

$S(n)$ with $n \geq 4$

$H(n)$ with $n \geq 5$, $n \neq 6$

$b(n)$ or $b'(n)$ with $n \geq 3$.

Now suppose that the G_0 - module G_1 is irreducible (i.e. suppose that $r = 1$ and hence that $G_1 = G_1^{(1)}$, $G_0 = G_0^{(1)}$).

If $G^{(1)}$ is isomorphic to one of the algebras

$spl(n,m)$ with $n > m \geq 1$

$spl(n,n) / K \cdot I_{2n}$ with $n \geq 3$

$osp(2,2r)$ with $r \geq 2$

$b(n)$ or $b'(n)$ with $n \geq 3$

H^{ξ} with a simple Lie algebra H

$S(n)$ with $n \geq 4$,

then proposition 3 of §5, lemma 6 of §1, and proposition 4 of §6 imply that $G^{(1)} = G$.

On the other hand, if $G^{(1)}$ is isomorphic to $H(n)$ with $n \geq 4$, $n \neq 6$, then we infer from proposition 8 of §6 that G is isomorphic to $H(n)$ or to $\tilde{H}(n)$. Note that this latter case includes the one in which $G^{(1)}$ is isomorphic to $spl(2,2) / K \cdot I_4$; in fact, we have already mentioned that the algebra $spl(2,2) / K \cdot I_4$ is isomorphic to $H(4)$.

Let us next consider the case where the G_0 - module G_1 is reducible. Then any two different of the G_0 - modules $G_1^{(s)}$, $1 \leq s \leq r$, are non-isomorphic.

In fact, suppose for example that there exists an isomorphism

$$g_1 : G_1^{(1)} \longrightarrow G_1^{(2)} \qquad (8.12)$$

of G_0 - modules. Then the algebras $G^{(1)}$ and $G^{(2)}$ are either both of the type 1) or both of the type 2).

Let us first assume that $G^{(1)}$ and $G^{(2)}$ are of the type 1). We know that the reductive Lie algebras $G_0^{(1)}$ and $G_0^{(2)}$ both have $\langle G_0 , G_0 \rangle$ as their semi-simple part. Since any two non-isomorphic algebras of the type 1) already differ in the semi-simple parts of their Lie algebras we conclude that $G^{(1)}$ and $G^{(2)}$ are isomorphic and that

$$G_0^{(1)} = G_0^{(2)} . \tag{8.13}$$

Let us define a linear mapping

$$g : G^{(1)} \longrightarrow G^{(2)} \tag{8.14,a}$$

by the requirement that

$$g(X) = X \qquad \text{if } X \in G_{-1} \tag{8.14,b}$$

$$g(Q) = Q \qquad \text{if } Q \in G_0^{(1)} \tag{8.14,c}$$

$$g(Y) = g_1(Y) \quad \text{if } Y \in G_1^{(1)} . \tag{8.14,d}$$

Moreover, let us define a new product $\langle \, , \, \rangle'$ on the Z - graded vector space $G^{(1)}$ by

$$\langle A , B \rangle' = g^{-1}(\langle g(A) , g(B) \rangle) \quad \text{for all } A , B \in G^{(1)} . \tag{8.15}$$

By definition $G^{(1)}$, equipped with the product $\langle \, , \, \rangle'$, is isomorphic to $G^{(2)}$ and hence is a simple Lie superalgebra. Obviously, we have

$$\langle Q , B \rangle' = \langle Q , B \rangle \quad \text{for all } Q \in G_0^{(1)} \text{ and } B \in G^{(1)} . \tag{8.16}$$

But then we infer from proposition 1 of §5 that there exists a constant $c \in K$ such that

$$\langle X , Y \rangle' = c \langle X , Y \rangle \quad \text{for all } X \in G_{-1} \text{ and } Y \in G_1^{(1)} , \tag{8.17}$$

which is to say that

$$\langle X , g_1(Y) \rangle = c \langle X , Y \rangle \quad \text{for all } X \in G_{-1} \text{ and } Y \in G_1^{(1)} . \tag{8.18}$$

The transitivity of G now implies that

$$g_1(Y) = c\,Y \quad \text{for all } Y \in G_1^{(1)} , \tag{8.19}$$

a contradiction.

We shall now suppose that $G^{(1)}$ and $G^{(2)}$ are of the type 2). In this case we have

$$G_0^{(1)} = \langle G_0 , G_0 \rangle = G_0^{(2)} . \tag{8.20}$$

Let us define a bilinear mapping

$$G_{-1} \times G_1^{(1)} \longrightarrow \langle G_0 , G_0 \rangle \tag{8.21,a}$$

by

$$(X,Y) \longrightarrow \langle X , g_1(Y) \rangle \quad \text{if } X \in G_{-1} \text{ and } Y \in G_1^{(1)} . \tag{8.21,b}$$

Obviously, this mapping is G_0 - invariant. It follows that there exists a constant $c \in K$ such that the relation (8.18) holds. (This is obvious if $G^{(1)} \simeq H^{\xi}$ with H a simple Lie algebra; for the remaining cases our assertion is known from the end of the proof of proposition 1' in §7.) Again we have arrived at a contradiction.

We now have to find out which of the algebras listed in 1) and 2) might possibly "combine" in a Z - graded Lie superalgebra G of the type under consideration. A careful examination of all possibilities reveals that only the following combinations are allowed :

S(n) and spl(n,1) , with $n \geq 3$

b'(4) and b'(4) .

(The reader should not be confused by the fact that in the second case the algebras $G^{(1)}$ and $G^{(2)}$ are isomorphic; in fact, this does *not* imply that the $\langle G_0 , G_0 \rangle$ - modules $G_1^{(1)}$ and $G_1^{(2)}$ are isomorphic, too.)

In the first case we conclude that G is isomorphic to W(n) (see §6, corollary to theorem 1), in the second case we deduce from proposition 8 of §6 that G is isomorphic to H(6) or to $\tilde{H}(6)$.

The theorem is proved.

Finally, as an easy consequence of the various results obtained thus far, we can prove the classification theorem for simple Lie superalgebras.

Theorem 2 (Kac)

We suppose that the field K is algebraically closed.

A simple Lie superalgebra is either a simple Lie algebra or else isomorphic to one of the following simple Lie superalgebras :

$spl(n,m)$ with $n, m \geq 1$; $n \neq m$

$spl(n,n) / K \cdot I_{2n}$ with $n \geq 2$

$osp(n,2r)$ with $n, r \geq 1$

$b(n)$ with $n \geq 3$

$d(n) / K \cdot I_{2n}$ with $n \geq 3$

$\Gamma(\sigma_1, \sigma_2, \sigma_3)$ with $\sigma_i \in K$, $\sigma_i \neq 0$, $\sigma_1 + \sigma_2 + \sigma_3 = 0$

Γ_2, Γ_3

$W(n)$ with $n \geq 2$

$S(n)$ with $n \geq 3$

$\tilde{S}(2r)$ with $r \geq 1$

$H(n)$ with $n \geq 4$.

Remark 4)

The following is a complete list of isomorphisms between the Lie superalgebras specified above :

$$spl(n,m) \simeq spl(m,n) \quad \text{for all } n, m \tag{8.22}$$

$$W(2) \simeq spl(2,1) \simeq osp(2,2) \tag{8.23}$$

$$S(3) \simeq b(3) \tag{8.24}$$

$$\tilde{S}(2) \simeq osp(1,2) \tag{8.25}$$

$$H(4) \simeq spl(2,2) / K \cdot I_4 \tag{8.26}$$

$$osp(4,2) \simeq \Gamma(-2,1,1) \; . \tag{8.27}$$

Finally, two algebras $\Gamma(\sigma_1, \sigma_2, \sigma_3)$ and $\Gamma(\sigma_1', \sigma_2', \sigma_3')$ are isomorphic if and only if there exist a non-zero element $\tau \in K$ as well as a permutation π

of {1,2,3} such that

$$\sigma_i' = \tau \cdot \sigma_{\pi i} \quad \text{for } i = 1,2,3 \ . \tag{8.28}$$

Proof

Let L be a simple Lie superalgebra; we may assume that

$$L_{\bar{1}} \neq \{0\} \ . \tag{8.29}$$

If L is classical, i.e. if the representation of $L_{\bar{0}}$ in $L_{\bar{1}}$ is completely reducible, then L is isomorphic to one of the Lie superalgebras listed in §5, theorem 1 . In particular, this remark applies to the case where the representation of $L_{\bar{0}}$ in $L_{\bar{1}}$ is irreducible.

Hence we now may assume that the representation of $L_{\bar{0}}$ in $L_{\bar{1}}$ is reducible. Let $L^0_{\bar{1}}$ be any non-zero maximal (proper) $L_{\bar{0}}$ - invariant subspace of $L_{\bar{1}}$; we define

$$L^0 = L_{\bar{0}} \oplus L^0_{\bar{1}} \ . \tag{8.30}$$

Evidently, L^0 is a graded subalgebra of L . Since L is simple this subalgebra uniquely determines a transitive filtration $(L^r)_{r \in Z}$ of L (see §1, lemma 7).

Consider the Z - graded Lie superalgebra gr L which is associated with this filtration. We know (see §1, lemma 8) that gr L is a transitive irreducible consistently Z - graded Lie superalgebra. Moreover, we deduce from the equations (1.72) that

$$L^1_{\bar{1}} = L^0_{\bar{1}} \neq \{0\} \tag{8.31}$$

and hence that

$$gr_1 L \neq \{0\} \ . \tag{8.32}$$

Thus the Z - graded Lie superalgebra gr L is isomorphic to one of the algebras which have been listed in theorem 1 ; we shall use the notation introduced therein.

If the center of the Lie algebra $gr_0 L$ is non-trivial, then the Lie superalgebras L and gr L are isomorphic (see §1, proposition 6). This case, therefore, can be settled immediately (in particular, we see that the algebra gr L cannot be of the type (IV)).

Consequently, we now may assume that the Lie algebra $gr_0 L$ is semi-simple. Let us first suppose that the Z - graded Lie superalgebra gr L is of the type (I) or (III). Then we have

$$L^r = \{0\} \quad \text{if } r \geqslant 2 . \tag{8.33}$$

This relation implies that the Lie algebras $L_{\bar{0}}$ and $gr_0 L$ are isomorphic (in fact, the equations (1.72) show that $L_{\bar{0}}^1 = \{0\}$). Consequently, $L_{\bar{0}}$ is a semi-simple Lie algebra and hence L is a classical simple Lie superalgebra. (In particular, gr L cannot be of the type (III).)

Thus we are left with the following cases :

a) The Z - graded Lie superalgebra gr L is isomorphic to S(n) , $n \geqslant 4$. If n is odd (resp. even) this implies that the Lie superalgebra L is isomorphic to S(n) (resp. to S(n) or to $\tilde{S}(n)$) (see §6, proposition 6).

b) The Z - graded Lie superalgebra gr L is isomorphic to H(n) or to $\tilde{H}(n)$ with $n \geqslant 4$. Then the Lie superalgebra L is isomorphic to gr L (see §6, proposition 9). But the Lie superalgebra $\tilde{H}(n)$ is not simple, hence the case gr L $\simeq \tilde{H}(n)$ cannot occur.

Remark 5)

Once the simple Lie superalgebras over an algebraically closed field are classified one can tackle the same problem over an arbitrary field. Kac [3] has reduced the classification of the simple Lie superalgebras over an arbitrary field to the corresponding problem for simple Lie algebras (and their irreducible representations); in particular, he has classified all simple real Lie superalgebras (see also chapter III, §3, n°3, remark 2)).

Chapter III A survey of some further developments

§ 1 Superderivations of Clifford algebras and Lie superalgebras

Recall that in chapter I, §1, example 4) we have defined, for an arbitrary superalgebra T, the Lie superalgebra $\mathfrak{D}(T)$ of superderivations of T. Restricting our attention to the case where T is finite-dimensional there are at least three types of superalgebras which might take the role of the algebra T, namely, the exterior algebras, the Clifford algebras, and the Lie superalgebras themselves. The first case has been investigated in §6 of chapter II, the second and third case will be discussed in the present paragraph.

1. Superderivations of a Clifford algebra

Let V be a vector space of dimension $n \geq 1$ and let Q be a non-degenerate quadratic form on V. It is well-known that the Clifford algebra C(Q) of Q [36] is an associative superalgebra; hence the Lie superalgebra $\mathfrak{D}(C(Q))$ of superderivations of C(Q) is well-defined.

Let $C(Q)^{\sim}$ be the Lie superalgebra associated with C(Q) (see chapter I, §1, example 2)) and let

$$C(Q)^{\sim} \longrightarrow \mathfrak{D}(C(Q)) \quad , \quad a \longrightarrow \hat{a} \tag{1.1}$$

be the canonical homomorphism of Lie superalgebras (see chapter I, §1, example 4)). Recall that, for every element $a \in C(Q)$, the superderivation $\hat{a}$ of C(Q) is defined by

$$\hat{a}(x) = \langle a , x \rangle \quad \text{for all } x \in C(Q) \; , \tag{1.2}$$

where $\langle \; , \; \rangle$ denotes the multiplication in $C(Q)^{\sim}$.

Proposition 1

The canonical homomorphism (1.1) is surjective and its kernel is equal to $K \cdot 1$.

Thus every superderivation of C(Q) is inner and we have

$$\dim \mathcal{D}(C(Q)) = 2^n - 1 . \tag{1.3}$$

The Lie superalgebra $\mathcal{D}(C(Q))$ is not simple; in fact, the commutator algebra A(Q) of $\mathcal{D}(C(Q))$ (see chapter I, §1, example 1)) has codimension 1 in $\mathcal{D}(C(Q))$, i.e.

$$\dim A(Q) = 2^n - 2 . \tag{1.4}$$

Proposition 2

Suppose that $\dim V \geqslant 4$. Then the commutator algebra A(Q) of $\mathcal{D}(C(Q))$ is simple.

Let us now assume that the field K is *algebraically closed*. Taking advantage of the well-known theorems on the structure of Clifford algebras we can determine the structure of $C(Q)^{\sim}$ and hence (see the proposition 1) the structure of $\mathcal{D}(C(Q))$ and of A(Q) . We have the following isomorphisms of Lie superalgebras :

If n is even, $n = 2m$, $m \geqslant 1$, then

$$C(Q)^{\sim} \simeq pl(2^{m-1}, 2^{m-1}) \tag{1.5,a}$$

$$A(Q) \simeq spl(2^{m-1}, 2^{m-1}) / K\cdot I_{2^m} ; \tag{1.5,b}$$

if n is odd, $n = 2m - 1$, $m \geqslant 1$, then

$$C(Q)^{\sim} \simeq L(2^{m-1}) \tag{1.6,a}$$

$$A(Q) \simeq d(2^{m-1}) / K\cdot I_{2^m} \tag{1.6,b}$$

(where the Lie superalgebra L(r) has been defined in chapter II, equation (4.41)).

2. Superderivations of a Lie superalgebra

Let L be a Lie superalgebra and let $\mathcal{D}(L)$ be the Lie superalgebra of superderivations of L . Recall (see chapter I, §1) that the adjoint representation of L is a homomorphism

$$\mathrm{ad} : L \longrightarrow \mathcal{D}(L) \tag{1.7}$$

of Lie superalgebras. Obviously, ad is injective if and only if the center of L is trivial. For any element $A \in L$, we call $\mathrm{ad}\,A$ the *inner* superderivation defined by A ; every element of $\mathcal{D}(L)$ which does not belong to $\mathrm{ad}\,L$ is called an *outer* superderivation.

By definition, a linear mapping D of L into itself is a superderivation of L if and only if

$$\langle D , \mathrm{ad}\,A \rangle = \mathrm{ad}\,D(A) \quad \text{for all } A \in L . \tag{1.8}$$

(The bracket on the left hand side denotes the multiplication in the general linear Lie superalgebra pl(L).) This equation shows that $\mathrm{ad}\,L$ is a graded ideal of $\mathcal{D}(L)$.

In complete analogy to the Lie algebra case we now can define the semi-direct product of two Lie superalgebras G and J , as follows. Let

$$d : G \longrightarrow \mathcal{D}(J) \tag{1.9}$$

be a homomorphism of Lie superalgebras. For convenience we shall write d_A instead of $d(A)$, for all $A \in G$. On the direct product $G \times J$ of the vector spaces G and J we define a Z_2-gradation by

$$(G \times J)_\alpha = G_\alpha \times J_\alpha \quad \text{if } \alpha \in Z_2 \tag{1.10}$$

as well as a multiplication $\langle\ ,\ \rangle$ by

$$\langle (A,B) , (A',B') \rangle = (\langle A , A' \rangle , \langle B , B' \rangle + d_A(B') - (-1)^{\alpha'\beta} d_{A'}(B)) \\ \text{for all } A \in G ,\ B \in J_\beta ,\ A' \in G_{\alpha'} ,\ B' \in J \ ;\ \alpha', \beta \in Z_2 . \tag{1.11}$$

It is easy to check that in this way $G \times J$ becomes a Lie superalgebra which is called the *semi-direct product* of G and J with respect to the homomorphism $d : G \longrightarrow \mathcal{D}(J)$. Evidently, the canonical injections of G and J into $G \times J$ as well as the canonical projection of $G \times J$ onto G are homomorphisms of Lie superalgebras.

Next we shall comment on the question of how the algebra of superderivations might actually be determined. Let L be a finite-dimensional Lie superalgebra. Recall (see chapter II, §3, proposition 5) that L does not have any outer superderivations at all if the Killing form of L is non-degenerate. In the general case the following remark turns out to be useful. Let S be a Levi factor of the Lie algebra $L_{\bar{0}}$. We use the adjoint representation of L to introduce on $\mathfrak{D}(L)$ the structure of an S-module. Since S is semi-simple and since ad L is an S-submodule of $\mathfrak{D}(L)$ there exists an S-invariant subspace W of $\mathfrak{D}(L)$ which is complementary to ad L. But ad L is an ideal of $\mathfrak{D}(L)$. It follows (see equation (1.8)) that

$$\operatorname{ad} D(Q) = 0 \quad \text{for all } D \in W \text{ and } Q \in S . \tag{1.12}$$

This equation means that D is an endomorphism of the S-module L, for every $D \in W$. In favourable cases this condition is sufficiently strong to enable the determination of $\mathfrak{D}(L)$.

The above method has been used by Kac [3] and, independently, by the author to prove the following proposition.

<u>Proposition 3</u>

1) The Lie superalgebras

   $spl(n,m)$ with $n, m \geq 1$, $n \neq m$

   $osp(n,2r)$ with $n, r \geq 1$

   $\Gamma(\sigma_1,\sigma_2,\sigma_3)$ with $\sigma_i \in K$, $\sigma_i \neq 0$, $\sigma_1+\sigma_2+\sigma_3 = 0$

   Γ_2, Γ_3 [37]

   $W(n)$ with $n \geq 2$

   $\tilde{S}(2r)$ with $r \geq 1$

do not have any outer superderivations.

2) The Lie superalgebras L which we are going to consider next do have outer superderivations. We shall give a Lie superalgebra $\hat{L}$ which is isomorphic to $\mathfrak{D}(L)$. In all cases there exists a natural embedding of L into $\hat{L}$ which, it is understood, corresponds to the embedding of L into

$\mathfrak{D}(L)$ by means of the adjoint representation.

a) For the following classical simple Lie superalgebras we find :

$$\mathfrak{D}(spl(n,n)/K\cdot I_{2n}) \simeq pl(n,n)/K\cdot I_{2n} \quad \text{if } n \geq 3 \,. \tag{1.13}$$

The Lie superalgebra $\mathfrak{D}(b(n))$, $n \geq 3$, is isomorphic to the Lie superalgebra defined by the equations (4.36) of chapter II.

$$\mathfrak{D}(d(n)/K\cdot I_{2n}) \simeq L(n)/K\cdot I_{2n} \quad \text{if } n \geq 3 \tag{1.14}$$

(see chapter II, equation (4.41)).

$$\mathfrak{D}(spl(2,2)/K\cdot I_4) \simeq \Gamma(\sigma,-\sigma,0) \tag{1.15}$$

for any non-zero element $\sigma \in K$.

b) The algebras $\mathfrak{D}(S(n))$ and $\mathfrak{D}(H(n))$ will be realized as subalgebras of $W(n)$. Let C be the element of $W_0(n)$ such that

$$\langle C , X \rangle = rX \quad \text{for all } X \in W_r(n) \text{ and } r \in Z \,. \tag{1.16}$$

Then

$$\mathfrak{D}(S(n)) \simeq S(n) \oplus K\cdot C \quad \text{if } n \geq 3 \tag{1.17}$$

$$\mathfrak{D}(H(n)) \simeq \tilde{H}(n) \oplus K\cdot C \quad \text{if } n \geq 5 \,. \tag{1.18}$$

Remarks

1) If G is one of the Z-graded Lie superalgebras $spl(n,n)/K\cdot I_{2n}$, $b(n)$, or $S(n)$, with $n \geq 3$ in all cases, then $\mathfrak{D}(G)$ is isomorphic to the Lie superalgebra G^Z which has been defined in chapter II, §1, n°3.

2) It is remarkable that a simple Lie superalgebra may have outer superderivations at all; in fact, this is contrary to what is known for simple Lie algebras.

§ 2 A FEW REMARKS ON NILPOTENT, SOLVABLE, AND SEMI-SIMPLE LIE SUPERALGEBRAS

Throughout this paragraph we shall assume that the Lie superalgebras are *finite-dimensional*.

1. Nilpotent and solvable Lie superalgebras

As in the Lie algebra case a Lie superalgebra is called *nilpotent* (resp. *solvable*) if the ideals in the lower (i.e. descending) central series (resp. in the derived series) vanish for sufficiently large indices.

Engel's theorem and its direct consequences remain valid, and the proof is the same as for Lie algebras [9,29].

Proposition 1

Let $V \neq \{0\}$ be a Z_2 - graded vector space and let L be a (finite-dimensional) graded subalgebra of pl(V) such that the elements of $L_{\bar{0}}$ and $L_{\bar{1}}$ are nilpotent. Then there exists a non-zero element $v \in V$ such that $X(v) = 0$ for all $X \in L$.

Corollary 1

A Lie superalgebra L is nilpotent if and only if $ad_L X$ is nilpotent for every homogeneous element X of L.

Corollary 2

The Lie superalgebra L in proposition 1 is nilpotent.

On the other hand, Lie's theorem does not necessarily hold for a solvable Lie superalgebra. Kac [3] has investigated the finite-dimensional irreducible graded representations of solvable Lie superalgebras. The following proposition contains two particularly neat results of his.

Proposition 2

a) A Lie superalgebra L is solvable if and only if its Lie algebra $L_{\bar{0}}$

is solvable.

b) Suppose that the field K is algebraically closed. All the finite-dimensional irreducible graded representations of a solvable Lie superalgebra L are one-dimensional if and only if $\langle L_{\bar{1}}, L_{\bar{1}} \rangle \subset \langle L_{\bar{0}}, L_{\bar{0}} \rangle$.

2. Semi-simple Lie superalgebras

For convenience we shall suppose in this section that the field K is *algebraically closed.*

Let us consider the following four properties of a Lie superalgebra L :

1) L does not contain any non-zero solvable graded ideals.

2) L is the direct product of finitely many simple Lie superalgebras.

3) The Killing form of L is non-degenerate.

4) All the finite-dimensional graded representations of L are completely reducible.

It is well-known that in the Lie algebra case these statements are mutually equivalent, they characterize the semi-simple Lie algebras. Nothing similar holds true for Lie superalgebras; in fact, it can be shown that in our list each statement is strictly stronger than the foregoing ones.

The Lie superalgebras of type 1) have been investigated by Kac [3]; he applies a method which has been used by Block [38] to study (among other things) the semi-simple Lie algebras over a field of prime characteristic. Regrettably, the resulting classification is far from being explicit. The Lie superalgebras of type 2) can be read off from theorem 2 in chapter II, §8. The corollaries to theorem 1 of chapter II, §3 and to theorem 1 of chapter II, §5 characterize the Lie superalgebras of type 3). Finally, the Lie superalgebras of type 4) will be discussed in section 1 of the subsequent paragraph. It turns out that the condition 4) is very restrictive; in fact, it rules out all simple Lie superalgebras except osp(1,2r) , $r \geq 1$ (and, of course, except the simple Lie algebras).

We conclude this paragraph by mentioning another negative result : Levi's radical splitting theorem does not hold for Lie superalgebras For example, it is obvious that spl(n,n) , $n \geqslant 2$, does not contain a graded subalgebra which is isomorphic to spl(n,n) / $K \cdot I_{2n}$.

§ 3 Finite-dimensional representations of simple Lie superalgebras

Throughout this paragraph we shall suppose that the field K is *algebraically closed* (in the last section K will be the field $\mathbb{C}$ of complex numbers). All Lie superalgebras are assumed to be *finite-dimensional* and all representations of Lie superalgebras are assumed to be *graded.*

1. Lie superalgebras all of whose finite-dimensional representations are completely reducible

As has already been stressed there exist simple Lie superalgebras some of whose finite-dimensional representations are not completely reducible. In fact, if a simple Lie superalgebra L has outer superderivations (see §1, proposition 3) then the canonical representation of L in the algebra of superderivations $\mathfrak{D}(L)$ is not completely reducible. One might still be optimistic and hope that "the non-completely-reducible representations are somehow exceptional and that they do not interfere with the normal representations which are completely reducible". But this is not the case; for example, there exists a typical (see section 2) irreducible 4 - dimensional representation ρ of spl(2,1) such that the tensor product of ρ with itself is not completely reducible [39]. Moreover, according to the following theorem by Djoković and Hochschild [40,41] it is really exceptional that a Lie superalgebra has all its finite-dimensional representations completely reducible.

Theorem 1

Let L be a Lie superalgebra over an algebraically closed field. All the finite-dimensional representations of L are completely reducible if and only if L is isomorphic to the direct product of a semi-simple Lie algebra with finitely many Lie superalgebras of the type osp(1,2r) , $r \geq 1$.

We shall indicate how this theorem has been proved. Till the end of this section *all vector spaces are assumed to be finite-dimensional.*

Let L be a Lie superalgebra such that all L - modules are completely reducible. For any $L_{\bar{0}}$ - module V we can construct the induced L - module

$$\bar{V} = U(L) \underset{o}{\otimes} V \tag{3.1}$$

where $\underset{o}{\otimes}$ indicates tensoring with respect to $U(L_{\bar{0}})$ (see chapter I, §4, n°1). Recall that if V is finite-dimensional (which we suppose), then so is $\bar{V}$. By assumption, $\bar{V}$ is completely reducible.

For a non-semi-simple Lie algebra $L_{\bar{0}}$ one can achieve by a judicious choice of the $L_{\bar{0}}$ - module V that the L - module $\bar{V}$ is not completely reducible. This implies :

A) The Lie algebra $L_{\bar{0}}$ is semi-simple.

On the other hand, we may consider K as a trivial $L_{\bar{0}}$ - module or L - module. Let

$$g : U(L) \underset{o}{\otimes} K \longrightarrow K \tag{3.2,a}$$

be the canonical homomorphism of L - modules which is defined by

$$g(X \otimes 1) = X_K(1) = \varepsilon(X) \quad \text{if } X \in U(L) \tag{3.2,b}$$

(see chapter I, §3, n°1, example 1)). Then there exists an L - submodule of $U(L) \underset{o}{\otimes} K$ which is complementary to kernel(g) . This means :

B) There exists an L - invariant element F in $U(L)_{\bar{0}} \underset{o}{\otimes} K$ such that $g(F) \neq 0$.

Conversely, let L be a Lie superalgebra such that the conditions A) and B) are satisfied. Then all L - modules are completely reducible.

An important consequence of B) is that

$$\langle U, U \rangle \neq 0 \quad \text{for all } U \in L_{\bar{1}},\ U \neq 0 . \tag{3.3}$$

As before we assume that L is a Lie superalgebra all of whose modules are completely reducible. Obviously, L is the direct product of simple Lie superalgebras which have the same property. The converse of this statement is also true, however, the proof is more difficult. Thus we may *assume that* L *is simple*. If $L_{\bar{1}} \neq \{0\}$ it follows that $L_{\bar{0}}$ is simple and that the representation of $L_{\bar{0}}$ in $L_{\bar{1}}$ is irreducible. But then we deduce from the relation (3.3) that L is isomorphic to one of the algebras osp(1,2r) , $r \geq 1$ (see lemma 2 of the appendix).

It remains to show that the representations of the Lie superalgebras osp(1,2r) really are completely reducible. This can be achieved by constructing an element F as described in the condition B) [41]; on the other hand, a suitable modification of the corresponding proof for semisimple Lie algebras (using the quadratic Casimir element) also yields the wanted result [32].

We conclude this section with the remark that Corwin [42] and Djoković [43] have developed a rather detailed representation theory for the Lie superalgebras osp(1,2r) .

2. Irreducible representations of simple Lie superalgebras

We shall next give a short review of Kac's work on the finite-dimensional irreducible representations of simple Lie superalgebras. His constructions are based on the theory of induced representations (see chapter I, §4, n^o1).

The so-called *basic* classical simple Lie superalgebras spl(n,m) with $n \neq m$, $spl(n,n)/K\cdot I_{2n}$, osp(n,2r) , $\Gamma(\sigma_1,\sigma_2,\sigma_3)$, Γ_2 , Γ_3 as well as the (f,d) algebras $d(n)/K\cdot I_{2n}$ can be treated by standard techniques [3]. Let L be one of these algebras. Choose a Cartan subalgebra h of $L_{\bar{0}}$ as well as a fundamental system of simple roots of L with respect to h . (For the basic algebras one can achieve that *only one simple root is odd.*) Let b be the corresponding "Borel subalgebra" of L ; of course, b contains also part of the odd subspace $L_{\bar{1}}$.

For any linear form $\lambda \in h^*$ there exists an obvious one-dimensional even b - module $W(\lambda)$ (the "nilpotent part" of b acts trivially). Consider the L - module $U(L) \otimes_{U(b)} W(\lambda)$ which is induced from $W(\lambda)$. This L - module contains a unique maximal (proper) L - submodule $I(\lambda)$. We define

$$V(\lambda) = (U(L) \otimes_{U(b)} W(\lambda)) / I(\lambda) \,. \qquad (3.4)$$

Then $V(\lambda)$ is an irreducible L - module with the highest weight λ . If

$\lambda_1, \lambda_2 \in h^*$, then the L-modules $V(\lambda_1)$ and $V(\lambda_2)$ are isomorphic if and only if $\lambda_1 = \lambda_2$.

Every finite-dimensional irreducible L-module V has a unique highest weight μ and V is isomorphic to $V(\mu)$ (possibly not until after a trivial change of the Z_2-gradation of V). The main task is then to find out which of the L-modules $V(\lambda)$ are finite-dimensional. This can in fact be done; let us mention, however, that for some algebras the corresponding conditions on λ look somewhat bewildering.

The simple Lie superalgebras b(n), W(n), S(n) and H(n) are treated by a different procedure [3]. Let G denote one of these algebras. Recall that G is a consistently Z-graded Lie superalgebra of the type

$$G = \bigoplus_{r \geq -1} G_r \quad . \tag{3.5}$$

We set

$$G^+ = \bigoplus_{r \geq 1} G_r \quad . \tag{3.6}$$

Let V be any finite-dimensional irreducible G-module. Then

$$V_0 = \{ v \in V \mid G^+(v) = \{0\} \} \tag{3.7}$$

is an irreducible G_0-submodule of V [44]. Moreover, two finite-dimensional irreducible G-modules V and V' are isomorphic if and only if the corresponding G_0-modules V_0 and V'_0 are isomorphic.

Conversely, let W be any finite-dimensional irreducible G_0-module. We define $G' = G_0 \oplus G^+$ and introduce on W the structure of a G'-module by setting $G^+(W) = \{0\}$. The induced G-module $U(G) \otimes_{U(G')} W$ is finite-dimensional and has an irreducible quotient module V such that the corresponding G_0-module V_0 is isomorphic to the G_0-module W.

In this way we have established a bijective correspondence between the finite-dimensional irreducible G-modules V and the finite-dimensional irreducible G_0-modules V_0.

Let us remark that the above procedure can also be applied to the Z-graded Lie superalgebras spl(n,m), spl(n,n)/$K \cdot I_{2n}$ and osp(2,2r).

The representations of the algebras $\tilde{S}(2r)$, $r \geq 2$, have not been discussed in [3].

In a later article [45] Kac has investigated in greater detail the finite-dimensional irreducible representations of the basic classical simple Lie superalgebras spl(n,m) with $n \neq m$, spl(n,n) / $K \cdot I_{2n}$, osp(n,2r), $\Gamma(\sigma_1,\sigma_2,\sigma_3)$, Γ_2, Γ_3. Let L denote one of these algebras. Kac has computed the character and the supercharacter of the so-called *typical* finite-dimensional irreducible L-modules. Without going into the details we remark that the typical L-modules are those which, in a sense, are in "general position". Kac has derived necessary and sufficient conditions for a finite-dimensional irreducible L-module V to be typical; a simple necessary condition is that

$$\dim V_{\bar{0}} = \dim V_{\bar{1}} , \tag{3.8}$$

provided that L is not isomorphic to one of the algebras osp(1,2r) (the finite-dimensional irreducible representations of these latter algebras are all typical) [46]. His discussion is based on Chevalley's theorem for Lie superalgebras (which, incidentally, only holds in a weakened form).

3. Generalized adjoint operations and star representations [47]

Throughout this section we shall suppose that K *is the field* $\mathbb{C}$ *of complex numbers*. All vector spaces (and, as before, all Lie superalgebras) are assumed to be *finite-dimensional*.

Let V be a Z_2-graded complex vector space and let $(\ |\ )$ be a non-degenerate hermitean form on V (linear in the second variable). We assume that $(\ |\ )$ is even, i.e. that the subspaces $V_{\bar{0}}$ and $V_{\bar{1}}$ are orthogonal with respect to $(\ |\ )$

$$(V_{\bar{0}} | V_{\bar{1}}) = \{0\} . \tag{3.9}$$

A Z_2-graded vector space V which is equipped with a non-degenerate even hermitean form will be called a Z_2-*graded hermitean vector space*, it will be called a Z_2 - *graded Hilbert space* if the hermitean form is positive definite.

For every linear mapping A of V into itself the adjoint of A (with respect to $(\ |\ )$) is denoted by A^+,

$$(A^+x|y) = (x|Ay) \quad \text{for all } x, y \in V . \tag{3.10}$$

The well-known rules for the adjoint operation $A \longrightarrow A^+$ in Hom(V) lead to the following definition.

Definition 1

Let L be a complex Lie superalgebra. An *adjoint operation* in L is an even semi-linear mapping $A \longrightarrow A^+$ of L into itself such that

$$\langle A, B\rangle^+ = \langle B^+, A^+\rangle \tag{3.11,a}$$

$$(A^+)^+ = A \tag{3.11,b}$$

for all $A, B \in L$.

For graded hermitean vector spaces V there exists a second natural (generalized) adjoint operation in Hom(V). Let $A \in \text{Hom}(V)_\alpha$, $\alpha \in Z_2$; we define the *superadjoint* $A^\ddagger$ of A by

$$(A^\ddagger x|y) = (-1)^{\alpha\xi}(x|Ay) \quad \text{for all } x \in V_\xi, \ y \in V;\ \xi \in Z_2 . \tag{3.12}$$

Through additivity, this definition is extended to all of Hom(V). Note that

$$(AB)^\ddagger = (-1)^{\alpha\beta} B^\ddagger A^\ddagger \tag{3.13}$$

for all $A \in \text{Hom}(V)_\alpha$, $B \in \text{Hom}(V)_\beta$; $\alpha, \beta \in Z_2$.

The rules which hold for the operation $A \longrightarrow A^\ddagger$ in Hom(V) suggest the following definition.

Definition 2

Let L be a complex Lie superalgebra. A *superadjoint operation* in L is an even semi-linear mapping $A \longrightarrow A^\ddagger$ of L into itself such that

$$\langle A, B\rangle^{\ddagger} = -\langle A^{\ddagger}, B^{\ddagger}\rangle \qquad (3.14,a)$$

$$(A^{\ddagger})^{\ddagger} = (-1)^{\alpha} A \qquad (3.14,b)$$

for all $A \in L_{\alpha}$, $B \in L$; $\alpha \in Z_2$.

Definition 3

Let L be a complex Lie superalgebra, equipped with an adjoint operation $A \longrightarrow A^{+}$ (resp. with a superadjoint operation $A \longrightarrow A^{\ddagger}$). Let V be a Z_2-graded hermitean vector space. A graded representation ρ of L in V is called a *star representation* (resp. a *superstar representation*) if

$$\rho(A^{+}) = \rho(A)^{+} \qquad (3.15)$$

$$(\text{ resp. } \quad \rho(A^{\ddagger}) = \rho(A)^{\ddagger} \quad) \qquad (3.16)$$

for all $A \in L$.

It is easy to see that every star or superstar representation of a Lie superalgebra in a graded *Hilbert space* is completely reducible. In view of the discussion in section 1 this result is quite welcome. A few elementary properties of star and superstar representations are given in [47]; the star and superstar representations of the algebras osp(1,2) and spl(2,1) have been investigated in [39].

Remark 1)

Let $\varepsilon \in \mathbb{C}$ be a square root of -1 . We define a sesquilinear form ψ on V by setting

$$\psi(x,y) = (x|y) \quad \text{if } x, y \in V_{\bar{0}} \qquad (3.17,a)$$

$$\psi(x,y) = \varepsilon\,(x|y) \quad \text{if } x, y \in V_{\bar{1}} \qquad (3.17,b)$$

$$\psi(x,y) = 0 \quad \text{if } x, y \in V \text{ are homogeneous of different degrees .} \qquad (3.17,c)$$

Then ψ is a non-degenerate even *superhermitean* form on V in the sense that

$$\overline{\psi(x,y)} = (-1)^{\xi\eta}\,\psi(y,x) \quad \text{for all } x \in V_{\xi}, \ y \in V_{\eta} \ ; \ \xi, \eta \in Z_2 \ . \qquad (3.18)$$

If $A \in Hom(V)_\alpha$, $\alpha \in Z_2$, we define an element $\tilde{A} \in Hom(V)_\alpha$ by

$$\psi(\tilde{A}x,y) = (-1)^{\alpha\xi}\psi(x,Ay) \quad \text{for all } x \in V_\xi , \ y \in V ; \ \xi \in Z_2 . \qquad (3.19)$$

As usual, we extend this definition through additivity to all of Hom(V). It is easy to check that $A \longrightarrow \tilde{A}$ is an even semi-linear mapping of pl(V) into itself such that

$$\langle A , B \rangle^{\sim} = -\langle \tilde{A} , \tilde{B} \rangle \qquad (3.20,a)$$

$$\tilde{\tilde{A}} = A \qquad (3.20,b)$$

for all $A , B \in pl(V)$.

On the other hand, we have

$$\tilde{A} = A^+ \quad \text{if } A \in Hom(V)_{\bar{0}} \qquad (3.21,a)$$

$$\tilde{A} = \varepsilon A^+ \quad \text{if } A \in Hom(V)_{\bar{1}} \qquad (3.21,b)$$

(where A^+ is defined with respect to (|)).

According to our general rules for dealing with Z_2 - graded objects it would be more adequate to work with ψ instead of (|). For example, if $A \longrightarrow \tilde{A}$ is an even semi-linear mapping of a Lie superalgebra L into itself which satisfies the relations (3.20), then

$$L^R = \{ A \in L \mid \tilde{A} = -A \} \qquad (3.22)$$

is a real form of L.

However, in view of the simple relationship (3.21) our choice to work with hermitean forms does not cause any trouble.

Finally, let us comment on how the adjoint and superadjoint operations in a Lie superalgebra may actually be determined. This will also shed some light on the question of "how natural" the definitions 1 and 2 are. We give one more definition.

Definition 4

Let L be a complex Lie superalgebra. A *generalized adjoint operation* in L is a bijective even semi-linear mapping σ of L onto itself which satisfies

$$\sigma(\langle Q , Q' \rangle) = - \langle \sigma(Q) , \sigma(Q') \rangle \quad (3.23,a)$$

$$\sigma \circ \sigma(Q) = Q \quad (3.23,b)$$

$$\sigma(\langle Q , U \rangle) = - \langle \sigma(Q) , \sigma(U) \rangle \quad (3.23,c)$$

for all $Q , Q' \in L_{\bar{0}}$ and $U \in L_{\bar{1}}$.

Note that the operations in the definitions 1 , 2 and in remark 1) are generalized adjoint operations in the sense of definition 4 .

The restriction of σ to $L_{\bar{0}}$ is an adjoint operation in the Lie algebra $L_{\bar{0}}$, thus

$$L^R_{\bar{0}} = \{ Q \in L_{\bar{0}} \mid \sigma(Q) = - Q \} \quad (3.24)$$

is a real form of the complex Lie algebra $L_{\bar{0}}$. Moreover, we deduce from equation (3.23,c) that the representation of $L_{\bar{0}}$ in $L_{\bar{1}}$ is equivalent to its complex conjugate representation (with respect to $L^R_{\bar{0}}$). Conversely, if we are given a real form $\hat{L}^R_{\bar{0}}$ of $L_{\bar{0}}$ such that the representation of $L_{\bar{0}}$ in $L_{\bar{1}}$ is equivalent to its complex conjugate representation (with respect to $\hat{L}^R_{\bar{0}}$), then there exists a generalized adjoint operation σ in L for which $L^R_{\bar{0}} = \hat{L}^R_{\bar{0}}$.

Now suppose that L is one of the classical simple Lie superalgebras except $\Gamma(\sigma_1,\sigma_2,\sigma_3)$ [48]. Then it follows from the proposition 1 of chapter II, §5 that there exists a non-zero complex number t such that

$$\sigma(\langle U , U' \rangle) = t \langle \sigma(U) , \sigma(U') \rangle \quad \text{for all } U , U' \in L_{\bar{1}} . \quad (3.25)$$

For simplicity, let us now disregard the case $L \simeq spl(2,2) / K \cdot I_4$, too. Then it is not difficult to see that the restriction of σ to $L_{\bar{1}}$ can be modified in such a way that σ becomes an adjoint or a superadjoint operation in L (in some cases, both of these "renormalizations" of σ are possible). For a detailed discussion the reader is referred to [47].

Remark 2)

The above results may be used to determine the real forms of the complex classical simple Lie superalgebras.

Appendix

In this appendix we shall collect our notational conventions for reductive Lie algebras. Furthermore, we shall discuss some classes of irreducible representations of semi-simple Lie algebras. For more details we refer the reader to the literature [9-12, 49].

Throughout the appendix we shall assume that the Lie algebras are *finite-dimensional* and that the field K is *algebraically closed*.

1. Notational conventions for reductive Lie algebras

Let g be a reductive Lie algebra. Then g is the direct product of its center h^0 with the derived algebra $g' = [g,g]$, the latter being semi-simple.

We choose a Cartan subalgebra h of g. It is well-known that

$$h = h^0 \times h' \tag{A.1}$$

where h' is a Cartan subalgebra of g'. The dual space h^* of h and the direct product $(h^0)^* \times (h')^*$ will be identified by means of the well-known canonical isomorphism.

Let ρ be any finite-dimensional completely reducible representation of g. It follows that the restriction of ρ to h is also completely reducible. For every element $\lambda \in h^*$ we define

$$V^\lambda = \{ v \in V \mid \rho(H)v = \lambda(H)v \text{ for all } H \in h \} . \tag{A.2}$$

Then we have

$$V = \bigoplus_{\lambda \in h^*} V^\lambda . \tag{A.3}$$

The linear forms $\lambda \in h^*$ with $V^\lambda \neq \{0\}$ are called the *weights* of the representation ρ (with respect to h). If λ is a weight of ρ, then V^λ is called the *weight space* associated with λ and any *non-zero* element of V^λ is called a *weight vector* belonging to (or associated with) the weight λ.

Let us apply these remarks to the adjoint representation of g. Since h is a Cartan subalgebra of g we have

$$g^0 = h . \tag{A.4}$$

The *non-zero* weights of the adjoint representation of g are called the *roots* of g (with respect to h). If α is a root of g then g^α is called the *root space* associated with α and any *non-zero* element of g^α is called a *root vector* belonging to the root α. We note that the roots of g vanish on the center h^0 of g ; on the canonical identification of $(h')^*$ with a subspace of h^* the roots of g with respect to h are just the roots of g' with respect to h'. It is well-known that

$$\dim g^\alpha = 1 \quad \text{for all roots } \alpha . \tag{A.5}$$

Let ϕ be any non-degenerate symmetric invariant bilinear form on g (such forms do exist). It is easy to see that h^0 and g' are orthogonal with respect to ϕ.

The invariance of ϕ implies that for all $\lambda, \mu \in h^*$

$$\phi(g^\lambda, g^\mu) = \{0\} \quad \text{if } \lambda + \mu \neq 0 ; \tag{A.6}$$

consequently, the restriction of ϕ to $g^\lambda \times g^{-\lambda}$ is non-degenerate.

In particular, the restriction of ϕ to h is non-degenerate. It follows that for every $\lambda \in h^*$ there exists a unique element $H_\lambda \in h$ such that

$$\lambda(H) = \phi(H_\lambda, H) \quad \text{for all } H \in h . \tag{A.7}$$

If λ is any element of h^* we have

$$[X, Y] = \phi(X,Y) H_\lambda \quad \text{for all } X \in g^\lambda \text{ and } Y \in g^{-\lambda} . \tag{A.8}$$

Frequently, it will be convenient to choose, for every root α of g, a root vector $E_\alpha \in g^\alpha$ such that

$$[E_{-\alpha}, E_\alpha] = H_\alpha \quad \text{for all roots } \alpha . \tag{A.9}$$

The restriction of ϕ to h induces a non-degenerate symmetric bilinear form $(\mid)$ on h^* which is given by

$$(\lambda|\mu) = \phi(H_\lambda, H_\mu) = \lambda(H_\mu) = \mu(H_\lambda) \quad \text{for all } \lambda, \mu \in h^* . \tag{A.10}$$

Obviously, the subspaces $(h^0)^*$ and $(h')^*$ of h^* are orthogonal with respect to $(\ |\ )$. Recall that $(\alpha|\alpha) \neq 0$ for all roots α of g ; moreover, if α and β are two roots of g then $2(\alpha|\beta)/(\alpha|\alpha)$ is an integer which does not depend on the choice of ϕ.

Let $\alpha_1, \dots, \alpha_n$ be a fundamental system of simple roots of g' with respect to h' (in particular, $\alpha_1, \dots, \alpha_n$ is a basis of $(h')^*$). The elements $\lambda_1, \dots, \lambda_n$ of $(h')^*$ which are defined by

$$2\,\frac{(\lambda_i|\alpha_k)}{(\alpha_k|\alpha_k)} = \delta_{ik} \quad ; \quad 1 \leq i, k \leq n \tag{A.11}$$

are called the *fundamental weights* of g' (with respect to h' and corresponding to the system $\alpha_1, \dots, \alpha_n$). The definition of the linear forms λ_i does not depend on the special choice of ϕ. Of course, $\lambda_1, \dots, \lambda_n$ is a basis of $(h')^*$.

Let us again identify $(h')^*$ with a subspace of h^* ; then $\alpha_1, \dots, \alpha_n$ is also called a fundamental system of simple roots of g with respect to h and $\lambda_1, \dots, \lambda_n$ are the corresponding fundamental weights of g .

Let ρ be a finite-dimensional irreducible representation of g and let ρ' be the restriction of ρ to g'. Then ρ' is also irreducible; moreover, the elements of h^0 are represented by scalar multiples of the identity. It follows that the weights of ρ are the elements of h^* of the form $\lambda = (\mu, \lambda')$ where μ is a fixed element of $(h^0)^*$ and where λ' runs through the weights of ρ'. The weight $\lambda = (\mu, \lambda')$ is called the highest (resp. lowest) weight of ρ if and only if λ' is the highest (resp. lowest) weight of ρ'.

These remarks should be sufficient to show how the general notions from the theory of semi-simple Lie algebras and their representations are generalized to the case where the Lie algebra is reductive.

2. Remarks on semi-simple Lie algebras and their representations

In the following g will denote a *semi-simple* Lie algebra. We choose some non-degenerate invariant bilinear form ϕ on g (for example the Killing form). Let h be a Cartan subalgebra of g, let $\alpha_1,\dots,\alpha_n$ be a fundamental system of simple roots of g with respect to h and let $\lambda_1,\dots,\lambda_n$ be the corresponding system of fundamental weights.

A linear form $\lambda \in h^*$ is called *dominant* if it has the form

$$\lambda = \sum_{i=1}^{n} c_i \lambda_i \tag{A.12}$$

where the c_i are integers, $c_i \geq 0$.

The finite-dimensional irreducible representations of g are characterized (up to equivalence) by their highest weight. An element $\lambda \in h^*$ is the highest weight of some finite-dimensional irreducible representation of g if and only if it is dominant. If λ is dominant the corresponding irreducible representation of g will be denoted by $\rho(\lambda)$.

For every root α of g we define a reflection S_α of h^* by

$$S_\alpha(\mu) = \mu - 2\frac{(\mu|\alpha)}{(\alpha|\alpha)}\alpha \quad \text{for all } \mu \in h^* . \tag{A.13}$$

Again, S_α is independent of the special choice of ϕ. If α runs through all roots of g the reflections S_α generate a finite group of isometries of h^* (with respect to (|)); this group is called the *Weyl group*. The following lemma is well-known:

Lemma 1

a) The Weyl group permutes the weights of any finite-dimensional representation of g.

b) For a simple Lie algebra the Weyl group operates transitively on the roots of equal length.

Suppose we are given a finite-dimensional irreducible representation ρ of g. If ρ is equivalent with its contragredient representation we call ρ *self-contragredient*. This is the case if and only if there exists a non-zero (and hence non-degenerate) invariant bilinear form ψ on the representation space of ρ. It is well-known that ψ (if it exists at all) is uniquely determined up to a factor. In particular, ψ is either symmetric or skew-symmetric; in the former (resp. latter) case the representation ρ is called *orthogonal* (resp. *symplectic*).

3. Special remarks on simple Lie algebras

We use the notation introduced in section 2 .

Let us first specify our convention for the enumeration of the vertices in the Dynkin diagrams of the simple Lie algebras (this is the convention chosen by Tits [49]).

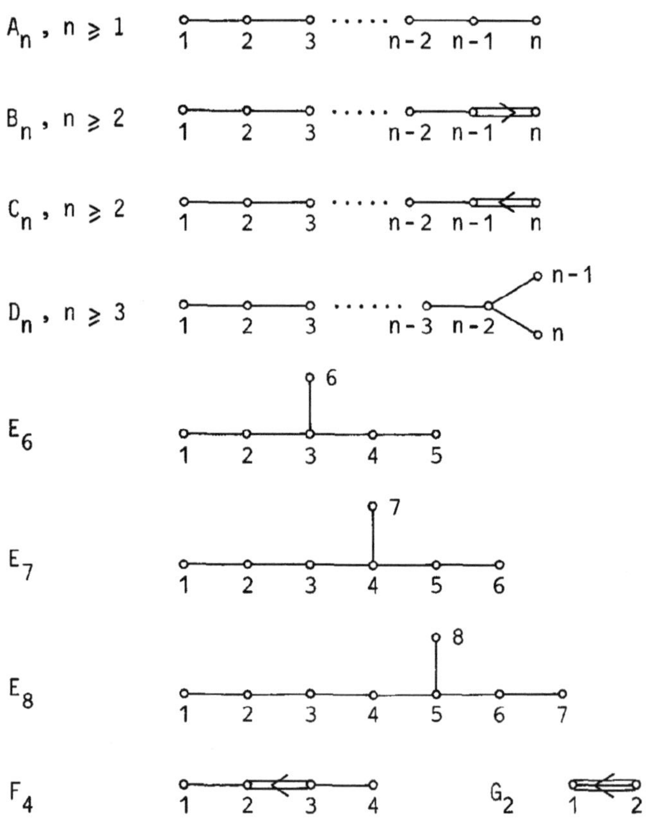

Figure : Enumeration of the vertices in the Dynkin diagrams of the simple Lie algebras

Let g be a *simple* Lie algebra. The representation $\rho(\lambda_1)$ of g is called *elementary*. In the cases of the Lie algebras A_n, B_n, C_n, D_n this is just the representation by which the algebra is usually defined, for G_2 it is the 7-dimensional representation in the space of traceless octonions.

We recall that for

$$A_n,\ n \geqslant 1 \qquad \overset{m}{\Lambda}\rho(\lambda_1) \simeq \rho(\lambda_m) \quad \text{if } 1 \leqslant m \leqslant n\ , \tag{A.14}$$

$$B_n,\ n \geqslant 2 \qquad \overset{m}{\Lambda}\rho(\lambda_1) \simeq \rho(\lambda_m) \quad \text{if } 1 \leqslant m \leqslant n-1 \tag{A.15,a}$$

$$\overset{n}{\Lambda}\rho(\lambda_1) \simeq \rho(2\lambda_n)\ , \tag{A.15,b}$$

$$C_n,\ n \geqslant 2 \qquad \overset{m}{\Lambda}\rho(\lambda_1) \simeq \rho(\lambda_m) \oplus \overset{m-2}{\Lambda}\rho(\lambda_1) \quad \text{if } 2 \leqslant m \leqslant n\ , \tag{A.16}$$

$$D_n,\ n \geqslant 3 \qquad \overset{m}{\Lambda}\rho(\lambda_1) \simeq \rho(\lambda_m) \quad \text{if } 1 \leqslant m \leqslant n-2 \tag{A.17,a}$$

$$\overset{n-1}{\Lambda}\rho(\lambda_1) \simeq \rho(\lambda_{n-1} + \lambda_n) \tag{A.17,b}$$

$$\overset{n}{\Lambda}\rho(\lambda_1) \simeq \rho(2\lambda_{n-1}) \oplus \rho(2\lambda_n)\ , \tag{A.17,c}$$

where, of course, $\overset{m}{\Lambda}\rho(\lambda_1)$ denotes the exterior product of m copies of the representation $\rho(\lambda_1)$.

The adjoint representation of g is irreducible. Hence g has a (unique) highest root which is, of course, dominant. If all the roots of g have the same length (which is the case for A_n, D_n, E_r) then no other dominant roots do exist (see lemma 1).

Suppose now that g has roots of different lengths (which is the case for B_n, C_n with $n \geqslant 2$, F_4, G_2). Then the highest root is a long root. Among the short roots there exists a unique highest one, and this is dominant. No further dominant roots do exist (again see lemma 1).

In table 1 we have listed the dominant roots of all simple Lie algebras.

algebra	highest root	additional dominant root
A_n , $n \geq 1$	$\lambda_1 + \lambda_n$	
B_n , $n \geq 3$	λ_2	λ_1
C_n , $n \geq 2$	$2\lambda_1$	λ_2
D_n , $n \geq 4$	λ_2	
E_6	λ_6	
E_7	λ_6	
E_8	λ_1	
F_4	λ_4	λ_1
G_2	λ_2	λ_1

Table 1 : Dominant roots of the simple Lie algebras

4. A technical lemma

The following lemma is due to Kac [3].

Lemma 2

We use the notation introduced in the sections 2 and 3 .

Let ρ be a faithful irreducible finite-dimensional representation of a non-zero semi-simple Lie algebra g . Let Ω be the set of all weights of ρ and let λ be the highest weight.

a) Suppose that 2λ is a root of g . Then

$$g \simeq C_n \quad , \quad \lambda = \lambda_1 \quad \text{for some } n \geq 1 .$$

b) Suppose that $\lambda - \tau$ is a root of g for all $\tau \in \Omega$, $\tau \neq \lambda$. Then the following cases are possible :

g	λ
A_n , $n \geq 2$	λ_1 , λ_n
C_n , $n \geq 1$	λ_1 .

c) Suppose that $\lambda - \tau$ is a root of g for all $\tau \in \Omega$, $\tau \neq \pm\lambda$. Then the following cases are possible :

g	λ
A_2	λ_1 , λ_2
A_3	$\lambda_1 , \lambda_2 , \lambda_3$
A_n , $n \geq 4$	λ_1 , λ_n
B_3	λ_1 , λ_3
B_n , $n \geq 4$	λ_1
C_1	$\lambda_1 , 2\lambda_1$
C_2	λ_1 , λ_2
C_n , $n \geq 3$	λ_1
D_4	$\lambda_1 , \lambda_3 , \lambda_4$
D_n , $n \geq 5$	λ_1
G_2	λ_1
$A_1 \times A_1$	(λ_1 , λ_1) .

Proof

Let μ be the lowest weight of ρ. It is well-known that $-\mu$ is the highest weight of the representation contragredient to ρ ; in particular, $-\mu$ is dominant. Recalling that ρ is faithful we conclude that 2λ or $\lambda - \mu$ can be a root of g only if g is simple. This remark applies to the cases a) and b).

We set

$$\lambda = \sum_{i=1}^{n} c_i \lambda_i \quad ; \quad -\mu = \sum_{i=1}^{n} d_i \lambda_i \tag{A.18}$$

with some integers $c_i , d_i \geq 0$. Since ρ is faithful we have

$$\sum_{i=1}^{n} c_i = \sum_{i=1}^{n} d_i \geq 1 \ . \tag{A.19}$$

a) By assumption 2λ is a dominant root of g . It is obvious from table 1 that this implies a).

b) By assumption $\lambda - \mu$ is a dominant root of g. In view of the inequality (A.19) our assertion again follows from table 1.

c) Obviously, we have $\lambda \neq \mu$. If also $\lambda \neq -\mu$, then $\lambda - \mu$ is a dominant root of g. This case has been settled in the proof of b).

Thus we now may assume that

$$-\mu = \lambda . \tag{A.20}$$

Let θ be the highest root of one of the simple components of g. Evidently, $\lambda - \theta$ is a weight of ρ. If $\lambda - \theta = \mu$ we conclude from equation (A.20) that $\theta = 2\lambda$; this case has been settled in a). Hence we may assume that $\lambda - \theta \neq \mu$. Then there exists a simple root α of g such that $\lambda - \theta - \alpha$ is a weight of ρ. This implies that $\lambda - \theta - \alpha = -\lambda$, for otherwise we would have $\lambda - \theta - \alpha \neq \pm\lambda$ and our assumption would imply that $\theta + \alpha = \lambda - (\lambda - \theta - \alpha)$ is a root of g. Thus we have shown that

$$2\lambda - \theta = \alpha \text{ is a simple root of } g . \tag{A.21}$$

Suppose now that g is not simple. Using once again the fact that ρ is faithful we conclude from (A.21) that $g \simeq A_1 \times A_1$ and that ρ is equivalent to the tensor product of the 2-dimensional fundamental representations of the two factors A_1.

Hence we now may assume that g is simple. Let us write

$$\theta = \sum_{i=1}^{n} t_i \lambda_i \tag{A.22}$$

with integers $t_i \geq 0$. Obviously, the equation (A.21) implies that

$$t_i = 2c_i - 2\frac{(\alpha|\alpha_i)}{(\alpha_i|\alpha_i)} \quad \text{if } 1 \leq i \leq n . \tag{A.23}$$

We have already mentioned that the numbers $2(\alpha|\alpha_i)/(\alpha_i|\alpha_i)$ are integers; moreover, it is known that $2(\alpha|\alpha_i)/(\alpha_i|\alpha_i) \leq -1$ if α and α_i belong to two neighbouring vertices of the Dynkin diagram. Hence we conclude from equation (A.23) that

$t_i \geq 1$ if α and α_i belong to two neighbouring vertices of the Dynkin diagram. (A.24)

We shall now distinguish three cases.

I) The simple root α does not belong to an extremal vertex of the Dynkin diagram.

According to (A.24) this implies that at least two of the coefficients t_i are greater than 1. Table 1 then shows that $g \simeq A_3$ and $\lambda = \lambda_2$.

II) The simple root α does belong to an extremal vertex of the Dynkin diagram; moreover, $\operatorname{rank} g \geq 3$.

Using (A.24) and table 1 as well as equation (A.21) we are led to the following cases :

g	λ
B_3	λ_1, λ_3
$B_n,\ n \geq 4$	λ_1
D_4	$\lambda_1, \lambda_3, \lambda_4$
$D_n,\ n \geq 5$	λ_1 .

III) $\operatorname{rank} g \leq 2$

In this case we find the following possibilities :

g	λ
C_1	$2\lambda_1$
C_2	λ_2
G_2	λ_1 .

The lemma is proved.

Remarks

1) The Lie algebras and representations listed in a), b), c) really do satisfy the corresponding conditions on their roots and weights.

2) The list in lemma 2,c) may be described as follows. It contains :

The algebras $sl(n)$, $n \geq 3$, combined with their standard representations and with the representations contragredient to the standard representations,

the algebras $sp(2r)$, $r \geq 1$, and the algebras $o(m)$, $m \geq 3$, combined with their standard representations,

the algebra $o(7)$ combined with the 8 - dimensional spin representation,

the algebra $o(8)$ combined with the two 8 - dimensional half-spin representations,

the algebra G_2 combined with the 7 - dimensional fundamental representation.

5. The index of a representation

In this section g denotes a *simple* Lie algebra (over the algebraically closed field K). All representations of g are assumed to be *finite-dimensional.*

Recall that with any representation ρ of g there is associated an invariant bilinear form ϕ_ρ on g which is defined by

$$\phi_\rho(Q, Q') = Tr(\rho(Q)\rho(Q')) \quad \text{for all } Q, Q' \in g . \tag{A.25}$$

The Killing form ϕ of g is obtained if we choose ρ to be the adjoint representation of g .

It is well-known that, since g is simple, any two invariant bilinear forms on g are proportional. Hence there exists an element $\ell_\rho \in K$ such that

$$\phi_\rho = \ell_\rho \phi . \tag{A.26}$$

The element ℓ_ρ is called the *index* of the representation ρ .

It is easy to prove the following (well-known) properties of the index.

a) If ρ is the direct sum of two subrepresentations ρ_1 and ρ_2 , then

$$\ell_\rho = \ell_{\rho_1} + \ell_{\rho_2} . \tag{A.27}$$

b) If ρ is the tensor product of two representations ρ_1 and ρ_2 whose dimensions are n_1 and n_2, respectively, then

$$\ell_\rho = n_2 \ell_{\rho_1} + n_1 \ell_{\rho_2} . \tag{A.28}$$

c) Let ρ be an n-dimensional representation of g.

α) The index of $\overset{m}{\Lambda}\rho$, $1 \leqslant m \leqslant n-1$, (the exterior product of m copies of ρ) is equal to $\binom{n-2}{m-1} \ell_\rho$.

β) The index of $S_m \rho$, $m \geqslant 1$, (the symmetric tensor product of m copies of ρ) is equal to $\binom{n+m}{m-1} \ell_\rho$.

d) The index of a representation ρ is equal to the index of the representation contragredient to ρ.

e) Suppose that the representation ρ is irreducible. Let C be the Casimir element which is constructed by means of the Killing form ϕ. Recall that C is an element of the center of the enveloping algebra $U(g)$ of g. Hence the image of C under the canonical extension of ρ to $U(g)$ has the form $c_\rho \cdot \mathrm{id}$ with some element $c_\rho \in K$. It follows that

$$\ell_\rho \cdot \dim g = c_\rho \cdot \dim \rho . \tag{A.29}$$

f) We use the notation introduced in the sections 1 and 2 and we assume that the bilinear form $(\mid)$ on h^* has been constructed by means of the Killing form ϕ of g. Suppose that ρ is an irreducible representation of g. If λ is the highest weight of ρ and if σ is half the sum of all positive roots (which is equal to the sum of the fundamental weights), then

$$c_\rho = (\lambda \mid \lambda + 2\sigma) . \tag{A.30}$$

Using the equations (A.27), (A.29) and (A.30) as well as Weyl's dimension formula we can calculate the index of any representation ρ of g. In particular, we conclude:

g) The index ℓ_ρ of any representation ρ of g is a positive rational number which is non-zero if and only if the representation ρ is faithful.

In the present work we need to know, for every simple Lie algebra g, which of the irreducible representations ρ of g have an index $\ell_\rho \leqslant 1$.

algebra	condition on the rank	highest weight of the representation	index ℓ_ρ
A_n	$n \geq 1$	λ_1 , λ_n	$\frac{1}{2(n+1)}$
	$n \geq 2$	$2\lambda_1 , 2\lambda_n$	$\frac{n+3}{2(n+1)}$
	$n \geq 2$	$\lambda_2 , \lambda_{n-1}$	$\frac{n-1}{2(n+1)}$
	$3 \leq n \leq 7$	$\lambda_3 , \lambda_{n-2}$	$\frac{(n-1)(n-2)}{4(n+1)}$
B_n	$n \geq 2$	λ_1	$\frac{1}{2n-1}$
	$2 \leq n \leq 6$	λ_n	$\frac{2^{n-3}}{2n-1}$
C_n	$n \geq 2$	λ_1	$\frac{1}{2(n+1)}$
	$n \geq 3$	λ_2	$\frac{n-1}{n+1}$
	$n = 2,3$	λ_n	$\frac{1}{2n(n+1)}\binom{2n}{n-1}$
D_n	$n \geq 4$	λ_1	$\frac{1}{2(n-1)}$
	$4 \leq n \leq 7$	$\lambda_{n-1} , \lambda_n$	$\frac{2^{n-5}}{n-1}$
E_6		λ_1 , λ_5	$\frac{1}{4}$
E_7		λ_1	$\frac{1}{3}$
F_4		λ_1	$\frac{1}{3}$
G_2		λ_1	$\frac{1}{4}$

Table 2 : All irreducible representations ρ with $0 < \ell_\rho < 1$

This problem has been solved in [50] (and, independently, by the author himself). The outcome is the following.

For any irreducible representation ρ of g the index ℓ_ρ and the dimension $\dim\rho$ of ρ satisfy:

$\ell_\rho < 1$ if and only if $\dim\rho < \dim g$

$\ell_\rho > 1$ if and only if $\dim\rho > \dim g$

$\ell_\rho = 1$ if and only if ρ is equivalent to the adjoint representation of g.

In table 2 we have listed, for every simple Lie algebra g, the highest weights of all non-trivial irreducible representations of g whose indices are strictly smaller than 1.

References and foot-notes

1 L. Corwin, Y. Ne'eman, and S. Sternberg, Rev. Mod. Phys. 47 (1975) 573

2 P. Fayet and S. Ferrara, Phys. Rep. 32C (1977) 249

3 V.G. Kac, Adv. Math. 26 (1977) 8
This work contains several references to earlier publications by Kac on the same subject.

4 V.G. Kac, Comm. Math. Phys. 53 (1977) 31

5 I. Kaplansky, Graded Lie algebras I, II, Univ. of Chicago report (1976)
See also
P.G.O. Freund and I. Kaplansky, J. Math. Phys. 17 (1976) 228

6 W. Nahm and M. Scheunert, J. Math. Phys. 17 (1976) 868

7 M. Scheunert, W. Nahm, and V. Rittenberg, J. Math. Phys. 17 (1976) 1626

8 M. Scheunert, W. Nahm, and V. Rittenberg, J. Math. Phys. 17 (1976) 1640

9 N. Bourbaki, Groupes et algèbres de Lie, chap. I ; Hermann, Paris (1960)

10 N. Bourbaki, Groupes et algèbres de Lie, chap. 2, 3 ; Hermann, Paris (1972)

11 N. Bourbaki, Groupes et algèbres de Lie, chap. IV - VI ; Hermann, Paris (1968)

12 N. Bourbaki, Groupes et algèbres de Lie, chap. 7, 8 ; Hermann, Paris (1975)

13 F.A. Berezin and D.A. Leites, Dokl. Akad. Nauk SSSR 224 (1975) 505 ; transl. Sov. Math. Dokl. 16 (1975) 1218

14 F.A. Berezin, Funct. Anal. Appl. 10 (1976) 70

15 F.A. Berezin, Institute of Theor. and Exp. Physics, ITEP - 66, 75 - 78 (5 parts), Moscow (1977)

16 B. Kostant, in "Differential geometrical methods in mathematical physics", Bonn 1975 ; Lecture Notes in Mathematics 570, Springer, Berlin (1977)

17 V. Rittenberg and M. Scheunert, J. Math. Phys. 19 (1978) 709

18 N. Bourbaki, Algèbre, chap. II, 3. edition; Hermann, Paris (1962)

19 N. Bourbaki, Algèbre, chap. III, new edition; Hermann, Paris (1971)

20 Let L be a Lie superalgebra and let G , G' be two subspaces of L . According to common usage we denote by $\langle G, G' \rangle$ the subspace of the vector space L which is generated by the elements of the form $\langle A, A' \rangle$ with $A \in G$ and $A' \in G'$.

21 L.E. Ross, Trans. Amer. Math. Soc. 120 (1965) 17

22 N. Bourbaki, Algèbre, chap. 8 ; Hermann, Paris (1958)

23 R.J. Blattner, Trans. Amer. Math. Soc. 144 (1969) 457

24 S. Sternberg, in "Differential geometrical methods in mathematical physics", Bonn 1975 ; Lecture Notes in Mathematics 570 , Springer, Berlin (1977)

25 V.W. Guillemin and S. Sternberg, Bull. Amer. Math. Soc. 70 (1964) 16

26 See S. Kobayashi and T. Nagano, J. Math. Mech. 14 (1965) 679

27 See [12], chap. 7 , §3

28 See [9], §6, proposition 6

29 N. Jacobson, Lie algebras; Interscience Publishers, Wiley, New York (1962)

30 See for example L. Michel, in "Group representations in mathematics and physics", Batelle Seattle 1969 Rencontres; Springer, Berlin, Heidelberg (1970)

31 M. Krämer, Comm. Alg. 3 (1975) 691 . Section 5 . Ausreduzierung einiger Tensorprodukte.
Note that his first formula for D_4 should read

$$S^2 \pi_2 = \pi_2^2 + \pi_1^2 + \pi_3^2 + \pi_4^2 + 1 .$$

32 D.Ž. Djoković, J. Pure Appl. Alg. 7 (1976) 217

33 "Algèbre de Cartan déployante" in the terminology of Bourbaki

34 The symmetric algebra of the vector space V* is denoted by Sym(V*) in

order to avoid a confusion with the Lie superalgebra S(V) .

35 See [9], §6, corollary 1 to theorem 5

36 N. Bourbaki, Algêbre, chap. 9 ; Hermann, Paris (1959)

37 We recall that the Lie superalgebras Γ_2 and Γ_3 have been introduced on the assumption that the field K is algebraically closed (see chapter II, §4, n°5). Note, however, that for any finite-dimensional Lie superalgebra L over any field K and for any extension field E of K , the Lie superalgebra $\mathcal{D}(E \otimes_K L)$ is canonically isomorphic to $E \otimes_K \mathcal{D}(L)$.

38 R.E. Block, Ann. Math. 90 (1969) 433

39 M. Scheunert, W. Nahm, and V. Rittenberg, J. Math. Phys. 18 (1977) 155

40 G. Hochschild, Illinois J. Math. 20 (1976) 107

41 D.Ž. Djokovič and G. Hochschild, Illinois J. Math. 20 (1976) 134

42 L. Corwin, Finite-dimensional representations of semi-simple graded Lie algebras, Rutgers Univ. report

43 D.Ž. Djokovič, J. Pure Appl. Alg. 9 (1976) 25

44 Note that the G_0 - module V_0 is Z_2 - graded. Of course, this gradation is trivial, the elements of V_0 being either all even or else all odd.

45 V.G. Kac, Comm. Alg. 5 (1977) 889
See also his recent MIT report on the same subject.

46 Let us remark that contrary to what is claimed in [45] a typical irreducible finite-dimensional L - module does not necessarily split in a finite-dimensional L - module.

47 M. Scheunert, W. Nahm, and V. Rittenberg, J. Math. Phys. 18 (1977) 146

48 Recall that osp(4,2) $\simeq \Gamma(-2,1,1)$; consequently, this algebra is also excluded.
Let σ be a generalized adjoint operation in $\Gamma(\sigma_1,\sigma_2,\sigma_3)$ and let π be the permutation of {1,2,3} such that

$$\sigma(sl(2)_i) = sl(2)_{\pi i} \quad \text{for } i = 1,2,3$$

where $sl(2)_i$ is the i. factor of $\Gamma(\sigma_1,\sigma_2,\sigma_3)_{\bar{0}} = sl(2) \times sl(2) \times sl(2)$. Then the relation (3.25) is fulfilled (with some non-zero element t $\in \mathbb{C}$) if and only if there exists a non-zero element $\tau \in \mathbb{C}$ such that

$$\bar{\sigma}_{\pi i} = \tau \cdot \sigma_i \quad \text{for } i = 1,2,3 .$$

49 J. Tits, Tabellen zu den einfachen Liegruppen und ihren Darstellungen; Lecture Notes in Mathematics 40 , Springer, Berlin (1967)

50 E.M. Andreev, E.B. Vinberg, and A.G. Elashvili, Funct. Anal. Appl. 1, n^{0}4 (1967) 3 ; transl. 1 (1967) 257

SUBJECT INDEX

Vol. 551: Algebraic K-Theory, Evanston 1976. Proceedings. Edited by M. R. Stein. XI, 409 pages. 1976.

Vol. 552: C. G. Gibson, K. Wirthmüller, A. A. du Plessis and E. J. N. Looijenga. Topological Stability of Smooth Mappings. V, 155 pages. 1976.

Vol. 553: M. Petrich, Categories of Algebraic Systems. Vector and Projective Spaces, Semigroups, Rings and Lattices. VIII, 217 pages. 1976.

Vol. 554: J. D. H. Smith, Mal'cev Varieties. VIII, 158 pages. 1976.

Vol. 555: M. Ishida, The Genus Fields of Algebraic Number Fields. VII, 116 pages. 1976.

Vol. 556: Approximation Theory. Bonn 1976. Proceedings. Edited by R. Schaback and K. Scherer. VII, 466 pages. 1976.

Vol. 557: W. Iberkleid and T. Petrie, Smooth S^1 Manifolds. III, 163 pages. 1976.

Vol. 558: B. Weisfeiler, On Construction and Identification of Graphs. XIV, 237 pages. 1976.

Vol. 559: J.-P. Caubet, Le Mouvement Brownien Relativiste. IX, 212 pages. 1976.

Vol. 560: Combinatorial Mathematics, IV, Proceedings 1975. Edited by L. R. A. Casse and W. D. Wallis. VII, 249 pages. 1976.

Vol. 561: Function Theoretic Methods for Partial Differential Equations. Darmstadt 1976. Proceedings. Edited by V. E. Meister, N. Weck and W. L. Wendland. XVIII, 520 pages. 1976.

Vol. 562: R. W. Goodman, Nilpotent Lie Groups: Structure and Applications to Analysis. X, 210 pages. 1976.

Vol. 563: Séminaire de Théorie du Potentiel. Paris, No. 2. Proceedings 1975–1976. Edited by F. Hirsch and G. Mokobodzki. VI, 292 pages. 1976.

Vol. 564: Ordinary and Partial Differential Equations, Dundee 1976. Proceedings. Edited by W. N. Everitt and B. D. Sleeman. XVIII, 551 pages. 1976.

Vol. 565: Turbulence and Navier Stokes Equations. Proceedings 1975. Edited by R. Temam. IX, 194 pages. 1976.

Vol. 566: Empirical Distributions and Processes. Oberwolfach 1976. Proceedings. Edited by P. Gaenssler and P. Révész. VII, 146 pages. 1976.

Vol. 567: Séminaire Bourbaki vol. 1975/76. Exposés 471–488. IV, 303 pages. 1977.

Vol. 568: R. E. Gaines and J. L. Mawhin, Coincidence Degree, and Nonlinear Differential Equations. V, 262 pages. 1977.

Vol. 569: Cohomologie Etale SGA 4½. Séminaire de Géométrie Algébrique du Bois-Marie. Edité par P. Deligne. V, 312 pages. 1977.

Vol. 570: Differential Geometrical Methods in Mathematical Physics, Bonn 1975. Proceedings. Edited by K. Bleuler and A. Reetz. VIII, 576 pages. 1977.

Vol. 571: Constructive Theory of Functions of Several Variables, Oberwolfach 1976. Proceedings. Edited by W. Schempp and K. Zeller. VI. 290 pages. 1977

Vol. 572: Sparse Matrix Techniques, Copenhagen 1976. Edited by V. A. Barker. V, 184 pages. 1977.

Vol. 573: Group Theory, Canberra 1975. Proceedings. Edited by R. A. Bryce, J. Cossey and M. F. Newman. VII, 146 pages. 1977.

Vol. 574: J. Moldestad, Computations in Higher Types. IV, 203 pages. 1977.

Vol. 575: K-Theory and Operator Algebras, Athens, Georgia 1975. Edited by B. B. Morrel and I. M. Singer. VI, 191 pages. 1977.

Vol. 576: V. S. Varadarajan, Harmonic Analysis on Real Reductive Groups. VI, 521 pages. 1977.

Vol. 577: J. P. May, E_∞ Ring Spaces and E_∞ Ring Spectra. IV, 268 pages. 1977.

Vol. 578: Séminaire Pierre Lelong (Analyse) Année 1975/76. Edité par P. Lelong. VI, 327 pages. 1977.

Vol. 579: Combinatoire et Représentation du Groupe Symétrique, Strasbourg 1976. Proceedings 1976. Edité par D. Foata. IV, 339 pages. 1977.

Vol. 580: C. Castaing and M. Valadier, Convex Analysis and Measurable Multifunctions. VIII, 278 pages. 1977.

Vol. 581: Séminaire de Probabilités XI, Université de Strasbourg. Proceedings 1975/1976. Edité par C. Dellacherie, P. A. Meyer et M. Weil. VI, 574 pages. 1977.

Vol. 582: J. M. G. Fell, Induced Representations and Banach *-Algebraic Bundles. IV, 349 pages. 1977.

Vol. 583: W. Hirsch, C. C. Pugh and M. Shub, Invariant Manifolds. IV, 149 pages. 1977.

Vol. 584: C. Brezinski, Accélération de la Convergence en Analyse Numérique. IV, 313 pages. 1977.

Vol. 585: T. A. Springer, Invariant Theory. VI, 112 pages. 1977.

Vol. 586: Séminaire d'Algèbre Paul Dubreil, Paris 1975–1976 (29ème Année). Edited by M. P. Malliavin. VI, 188 pages. 1977.

Vol. 587: Non-Commutative Harmonic Analysis. Proceedings 1976. Edited by J. Carmona and M. Vergne. IV, 240 pages. 1977.

Vol. 588: P. Molino, Théorie des G-Structures: Le Problème d'Equivalence. VI, 163 pages. 1977.

Vol. 589: Cohomologie l-adique et Fonctions L. Séminaire de Géométrie Algébrique du Bois-Marie 1965–66, SGA 5. Edité par L. Illusie. XII, 484 pages. 1977.

Vol. 590: H. Matsumoto, Analyse Harmonique dans les Systèmes de Tits Bornologiques de Type Affine. IV, 219 pages. 1977.

Vol. 591: G. A. Anderson, Surgery with Coefficients. VIII, 157 pages. 1977.

Vol. 592: D. Voigt, Induzierte Darstellungen in der Theorie der endlichen, algebraischen Gruppen. V, 413 Seiten. 1977.

Vol. 593: K. Barbey and H. König, Abstract Analytic Function Theory and Hardy Algebras. VIII, 260 pages. 1977.

Vol. 594: Singular Perturbations and Boundary Layer Theory, Lyon 1976. Edited by C. M. Brauner, B. Gay, and J. Mathieu. VIII, 539 pages. 1977.

Vol. 595: W. Hazod, Stetige Faltungshalbgruppen von Wahrscheinlichkeitsmaßen und erzeugende Distributionen. XIII, 157 Seiten. 1977.

Vol. 596: K. Deimling, Ordinary Differential Equations in Banach Spaces. VI, 137 pages. 1977.

Vol. 597: Geometry and Topology, Rio de Janeiro, July 1976. Proceedings. Edited by J. Palis and M. do Carmo. VI, 866 pages. 1977.

Vol. 598: J. Hoffmann-Jørgensen, T. M. Liggett et J. Neveu, Ecole d'Eté de Probabilités de Saint-Flour VI – 1976. Edité par P.-L. Hennequin. XII, 447 pages. 1977.

Vol. 599: Complex Analysis, Kentucky 1976. Proceedings. Edited by J. D. Buckholtz and T. J. Suffridge. X, 159 pages. 1977.

Vol. 600: W. Stoll, Value Distribution on Parabolic Spaces. VIII, 216 pages. 1977.

Vol. 601: Modular Functions of one Variable V, Bonn 1976. Proceedings. Edited by J.-P. Serre and D. B. Zagier. VI, 294 pages. 1977.

Vol. 602: J. P. Brezin, Harmonic Analysis on Compact Solvmanifolds. VIII, 179 pages. 1977.

Vol. 603: B. Moishezon, Complex Surfaces and Connected Sums of Complex Projective Planes. IV, 234 pages. 1977.

Vol. 604: Banach Spaces of Analytic Functions, Kent, Ohio 1976. Proceedings. Edited by J. Baker, C. Cleaver and Joseph Diestel. VI, 141 pages. 1977.

Vol. 605: Sario et al., Classification Theory of Riemannian Manifolds. XX, 498 pages. 1977.

Vol. 606: Mathematical Aspects of Finite Element Methods. Proceedings 1975. Edited by I. Galligani and E. Magenes. VI, 362 pages. 1977.

Vol. 607: M. Métivier, Reelle und Vektorwertige Quasimartingale und die Theorie der Stochastischen Integration. X, 310 Seiten. 1977.

Vol. 608: Bigard et al., Groupes et Anneaux Réticulés. XIV, 334 pages. 1977.

Vol. 609: General Topology and Its Relations to Modern Analysis and Algebra IV. Proceedings 1976. Edited by J. Novák. XVIII, 225 pages. 1977.

Vol. 610: G. Jensen, Higher Order Contact of Submanifolds of Homogeneous Spaces. XII, 154 pages. 1977.

Vol. 611: M. Makkai and G. E. Reyes, First Order Categorical Logic. VIII, 301 pages. 1977.

Vol. 612: E. M. Kleinberg, Infinitary Combinatorics and the Axiom of Determinateness. VIII, 150 pages. 1977.

Vol. 613: E. Behrends et al., L^p-Structure in Real Banach Spaces. X, 108 pages. 1977.

Vol. 614: H. Yanagihara, Theory of Hopf Algebras Attached to Group Schemes. VIII, 308 pages. 1977.

Vol. 615: Turbulence Seminar, Proceedings 1976/77. Edited by P. Bernard and T. Ratiu. VI, 155 pages. 1977.

Vol. 616: Abelian Group Theory, 2nd New Mexico State University Conference, 1976. Proceedings. Edited by D. Arnold, R. Hunter and E. Walker. X, 423 pages. 1977.

Vol. 617: K. J. Devlin, The Axiom of Constructibility: A Guide for the Mathematician. VIII, 96 pages. 1977.

Vol. 618: I. I. Hirschman, Jr. and D. E. Hughes, Extreme Eigen Values of Toeplitz Operators. VI, 145 pages. 1977.

Vol. 619: Set Theory and Hierarchy Theory V, Bierutowice 1976. Edited by A. Lachlan, M. Srebrny, and A. Zarach. VIII, 358 pages. 1977.

Vol. 620: H. Popp, Moduli Theory and Classification Theory of Algebraic Varieties. VIII, 189 pages. 1977.

Vol. 621: Kauffman et al., The Deficiency Index Problem. VI, 112 pages. 1977.

Vol. 622: Combinatorial Mathematics V, Melbourne 1976. Proceedings. Edited by C. Little. VIII, 213 pages. 1977.

Vol. 623: I. Erdelyi and R. Lange, Spectral Decompositions on Banach Spaces. VIII, 122 pages. 1977.

Vol. 624: Y. Guivarc'h et al., Marches Aléatoires sur les Groupes de Lie. VIII, 292 pages. 1977.

Vol. 625: J. P. Alexander et al., Odd Order Group Actions and Witt Classification of Innerproducts. IV, 202 pages. 1977.

Vol. 626: Number Theory Day, New York 1976. Proceedings. Edited by M. B. Nathanson. VI, 241 pages. 1977.

Vol. 627: Modular Functions of One Variable VI, Bonn 1976. Proceedings. Edited by J.-P. Serre and D. B. Zagier. VI, 339 pages. 1977.

Vol. 628: H. J. Baues, Obstruction Theory on the Homotopy Classification of Maps. XII, 387 pages. 1977.

Vol. 629: W.A. Coppel, Dichotomies in Stability Theory. VI, 98 pages. 1978.

Vol. 630: Numerical Analysis, Proceedings, Biennial Conference, Dundee 1977. Edited by G. A. Watson. XII, 199 pages. 1978.

Vol. 631: Numerical Treatment of Differential Equations. Proceedings 1976. Edited by R. Bulirsch, R. D. Grigorieff, and J. Schröder. X, 219 pages. 1978.

Vol. 632: J.-F. Boutot, Schéma de Picard Local. X, 165 pages. 1978.

Vol. 633: N. R. Coleff and M. E. Herrera, Les Courants Résiduels Associés à une Forme Méromorphe. X, 211 pages. 1978.

Vol. 634: H. Kurke et al., Die Approximationseigenschaft lokaler Ringe. IV, 204 Seiten. 1978.

Vol. 635: T. Y. Lam, Serre's Conjecture. XVI, 227 pages. 1978.

Vol. 636: Journées de Statistique des Processus Stochastiques, Grenoble 1977, Proceedings. Edité par Didier Dacunha-Castelle et Bernard Van Cutsem. VII, 202 pages. 1978.

Vol. 637: W. B. Jurkat, Meromorphe Differentialgleichungen. VII, 194 Seiten. 1978.

Vol. 638: P. Shanahan, The Atiyah-Singer Index Theorem, An Introduction. V, 224 pages. 1978.

Vol. 639: N. Adasch et al., Topological Vector Spaces. V, 125 pages. 1978.

Vol. 640: J. L. Dupont, Curvature and Characteristic Classes. X, 175 pages. 1978.

Vol. 641: Séminaire d'Algèbre Paul Dubreil, Proceedings Paris 1976–1977. Edité par M. P. Malliavin. IV, 367 pages. 1978.

Vol. 642: Theory and Applications of Graphs, Proceedings, Michigan 1976. Edited by Y. Alavi and D. R. Lick. XIV, 635 pages. 1978.

Vol. 643: M. Davis, Multiaxial Actions on Manifolds. VI, 141 pages. 1978.

Vol. 644: Vector Space Measures and Applications I, Proceedings 1977. Edited by R. M. Aron and S. Dineen. VIII, 451 pages. 1978.

Vol. 645: Vector Space Measures and Applications II, Proceedings 1977. Edited by R. M. Aron and S. Dineen. VIII, 218 pages. 1978.

Vol. 646: O. Tammi, Extremum Problems for Bounded Univalent Functions. VIII, 313 pages. 1978.

Vol. 647: L. J. Ratliff, Jr., Chain Conjectures in Ring Theory. VIII, 133 pages. 1978.

Vol. 648: Nonlinear Partial Differential Equations and Applications, Proceedings, Indiana 1976–1977. Edited by J. M. Chadam. VI, 206 pages. 1978.

Vol. 649: Séminaire de Probabilités XII, Proceedings, Strasbourg, 1976–1977. Edité par C. Dellacherie, P. A. Meyer et M. Weil. VIII, 805 pages. 1978.

Vol. 650: C*-Algebras and Applications to Physics. Proceedings 1977. Edited by H. Araki and R. V. Kadison. V, 192 pages. 1978.

Vol. 651: P. W. Michor, Functors and Categories of Banach Spaces. VI, 99 pages. 1978.

Vol. 652: Differential Topology, Foliations and Gelfand-Fuks-Cohomology, Proceedings 1976. Edited by P. A. Schweitzer. XIV, 252 pages. 1978.

Vol. 653: Locally Interacting Systems and Their Application in Biology. Proceedings, 1976. Edited by R. L. Dobrushin, V. I. Kryukov and A. L. Toom. XI, 202 pages. 1978.

Vol. 654: J. P. Buhler, Icosahedral Galois Representations. III, 143 pages. 1978.

Vol. 655: R. Baeza, Quadratic Forms Over Semilocal Rings. VI, 199 pages. 1978.

Vol. 656: Probability Theory on Vector Spaces. Proceedings, 1977. Edited by A. Weron. VIII, 274 pages. 1978.

Vol. 657: Geometric Applications of Homotopy Theory I, Proceedings 1977. Edited by M. G. Barratt and M. E. Mahowald. VIII, 459 pages. 1978.

Vol. 658: Geometric Applications of Homotopy Theory II, Proceedings 1977. Edited by M. G. Barratt and M. E. Mahowald. VIII, 487 pages. 1978.

Vol. 659: Bruckner, Differentiation of Real Functions. X, 247 pages. 1978.

Vol. 660: Equations aux Dérivée Partielles. Proceedings, 1977. Edité par Pham The Lai. VI, 216 pages. 1978.

Vol. 661: P. T. Johnstone, R. Paré, R. D. Rosebrugh, D. Schumacher, R. J. Wood, and G. C. Wraith, Indexed Categories and Their Applications. VII, 260 pages. 1978.

Vol. 662: Akin, The Metric Theory of Banach Manifolds. XIX, 306 pages. 1978.

Vol. 663: J. F. Berglund, H. D. Junghenn, P. Milnes, Compact Right Topological Semigroups and Generalizations of Almost Periodicity. X, 243 pages. 1978.

Vol. 664: Algebraic and Geometric Topology, Proceedings, 1977. Edited by K. C. Millett. XI, 240 pages. 1978.

Vol. 665: Journées d'Analyse Non Linéaire. Proceedings, 1977. Edité par P. Bénilan et J. Robert. VIII, 256 pages. 1978.

Vol. 666: B. Beauzamy, Espaces d'Interpolation Réels: Topologie et Géometrie. X, 104 pages. 1978.

Vol. 667: J. Gilewicz, Approximants de Padé. XIV, 511 pages. 1978.

Vol. 668: The Structure of Attractors in Dynamical Systems. Proceedings, 1977. Edited by J. C. Martin, N. G. Markley and W. Perrizo. VI, 264 pages. 1978.

Vol. 669: Higher Set Theory. Proceedings, 1977. Edited by G. H. Müller and D. S. Scott. XII, 476 pages. 1978.

MIX
Papier aus verantwortungsvollen Quellen
Paper from responsible sources
FSC® C105338

If you have any concerns about our products,
you can contact us on
ProductSafety@springernature.com

In case Publisher is established outside the EU,
the EU authorized representative is:
Springer Nature Customer Service Center GmbH
Europaplatz 3, 69115 Heidelberg, Germany

Printed by Libri Plureos GmbH
in Hamburg, Germany